SURVEYING YOUR LAND

SURVEYING YOUR LAND

A Common-Sense Guide to Surveys, Deeds, and Title Searches

Charles E. Lawson, R.L.S.

The Countryman Press

WOODSTOCK, VERMONT

Surveying Your Land was originally published in 1983 by the Stephen Greene Press as *Surveys, Deeds, and Title Searches* (ISBN 0-8289-0505-3).

Library of Congress Cataloging-in-Publication Data
Lawson, Charles E.
Surveying your land: a common-sense guide to surveys, deeds, and title searches / Charles E. Lawson.
p. cm.
Reprint. Originally published: Surveys, deeds, and title searches. Brattleboro, Vt. : Stephen Greene Press. c1983.
Includes index
ISBN 0-88150-180-8
1. Surveying—Law and legislation—United States—Popular works. 2. Deeds—United States—Popular works. 3. Title examination—United States—Popular works. I. Lawson, Charles E. Surveys, deeds, and title searches. II. Title.
KF683.Z9L38 1990
346.7304'38—dc20
[347.306438] 90-1998
CIP

Published by The Countryman Press, Inc.
Woodstock, Vermont 05091

Cover design by James Brisson

Figure 40A courtesy of Topcon Instrument Corporation of America;
Figure 42 by James Brisson; all other illustrations by the author.

Printed in the United States of America
10 9 8 7 6 5 4 3 2 1

Contents

Preface vii
Acknowledgments ix

Chapter 1 *Why a Survey?* 1
Chapter 2 *How a Property Survey is Made* 9
Chapter 3 *Aerial Photos—Tax Maps* 21
Chapter 4 *Researching a Parcel of Land* 29
Chapter 5 *Mathematical Closure* 39
Chapter 6 *Deeds and Deed Descriptions* 49
Chapter 7 *Marking Your Land* 65
Chapter 8 *Changing Corner Markers* 70
Chapter 9 *Lawyers and Legal Considerations, Including Title Search* 75
Chapter 10 *Assessing and Taxing Property* 85
Chapter 11 *Mortgage Inspections* 91
Chapter 12 *Surveying—Yesterday, Today, and Tomorrow* 93

Appendix A *Glossary* 108
Appendix B *Conversions* 112
Appendix C *Landowner's Checklists* 113

Index 116

Preface

This edition of *Surveying Your Land* (previously, *Surveys, Deeds and Title Searches,*) has been completely updated and should be a valuable handbook not only for new or potential landowners, but—as I have discovered from previous editions—for attorneys, building inspectors, assessors, town planners, and land surveyors. Revisions have been made and new material added—most notably on mortgage inspections, surveying by satellite, and computerized mapping—with all of these people in mind. I want to emphasize however, that this book is still intended, above all, for the individual land purchaser—the one who must make sure that all necessary steps have been taken. To leave these essential details to the bank, the real estate agency, or the lawyer writing the deed is irresponsible—and could be disastrous—considering the prices of land and the demand for it today.

Even where land prices are not yet inflated, it is still necessary for landowners to make sure that their property is correctly surveyed and recorded. As towns continue to upgrade their systems of mapping for tax purposes, it becomes increasingly important for landowners to check those maps to ensure that they are being taxed for their land only. Such vigilance is especially worthwhile in the case of property along roads, as it is susceptible to being mapped and taxed as potential house lots—with resulting high taxation. Such tax mapping is also of particular concern to citizens involved in planning, as classification of roadside property as house sites contributes to growth and strip development around our towns—something that planning boards have generally fought hard to prevent.

Our property is thus important not only to our personal welfare, but to that of our communities. In the final analysis, our property is that which is defined by a properly done survey. Just what a surveyor does is not widely and clearly understood. The more jobs I undertake, the more evident it becomes that public education is an important element in practicing the surveyor's profession.

I hope this book will shed some light on what a surveyor does and why it is important. Some readers may note that more detailed data could be provided, but my intention is not to write a textbook but to provide a book that will provide the landowner or purchaser with essential information.

Charles E. Lawson, R.L.S.
North Berwick, Maine

Acknowledgments

The author would like to express his appreciation to all who assisted in putting this book together, especially: Daniel Blanchette of Eliot, Maine, who assisted immensely in describing assessing methods; R.L. Reagan, C.E., and Jim Godfrey, R.L.S., of Houston, Texas, for maps and deeds of the Rio Grande area; the U.S. Department of the Interior, Geological Survey, National Cartographic Information Center, Reston, Virginia; the U.S. Department of Commerce, National Oceanic and Atmospheric Administration, Rockville, Maryland; Col. William Baugh, USAF (Ret.), Director of Public Affairs, 2nd Space Wing, Falcon Air Force Base, Colorado, and Donald Garrold, R.L.S., of Searsport, Maine, for their assistance in updating the material on the Global Positioning System; the Ottauquechee Regional Planning and Development Commission, Woodstock, Vermont, for assisting with the material on the Geographic Information System; and last, but not least, my wife, who has stood patiently by as I rambled on about surveying and why it should be done.

Chapter 1 *Why a Survey?*

What would be your reaction if you were told that there is a discrepancy between the boundary markers of land you own and your deed to that land? Or, what would your reaction be if you were told that there is a fair chance that the land you contemplate buying will have a deed that does not describe the boundary markers accurately enough to locate it precisely on the spot where you think it is? Right now parcels of land are being purchased along with complete surveys and with complete plans signed, sealed, and filed at the registry of deeds; yet unless the boundaries are validated, there can be trouble for the owner.

How does all this happen?

Age-old Problem

Problems with property boundaries have plagued man since he first began settling in one place. The Bible gives hints of this. In Deuteronomy 19:14 we read, "Thou shalt not remove thy neighbor's landmark," and later, in verse 27:17, it says, "Cursed be he that removeth his neighbor's landmark, and all the people shall say Amen." And in Proverbs 22:28 we find, "Remove not the ancient landmarks, which thy fathers have set."

Early Settlements

During America's first years of settlement, little was done to put order into the arrangement or layout of the land. Settlers came, squatted, and marked off boundaries of a parcel by referring to a particular brook, a line of trees, or a ridge. The deeds and plans of these parcels were made up accordingly by trying to describe these boundaries. Anyone looking at a map of the United States will see

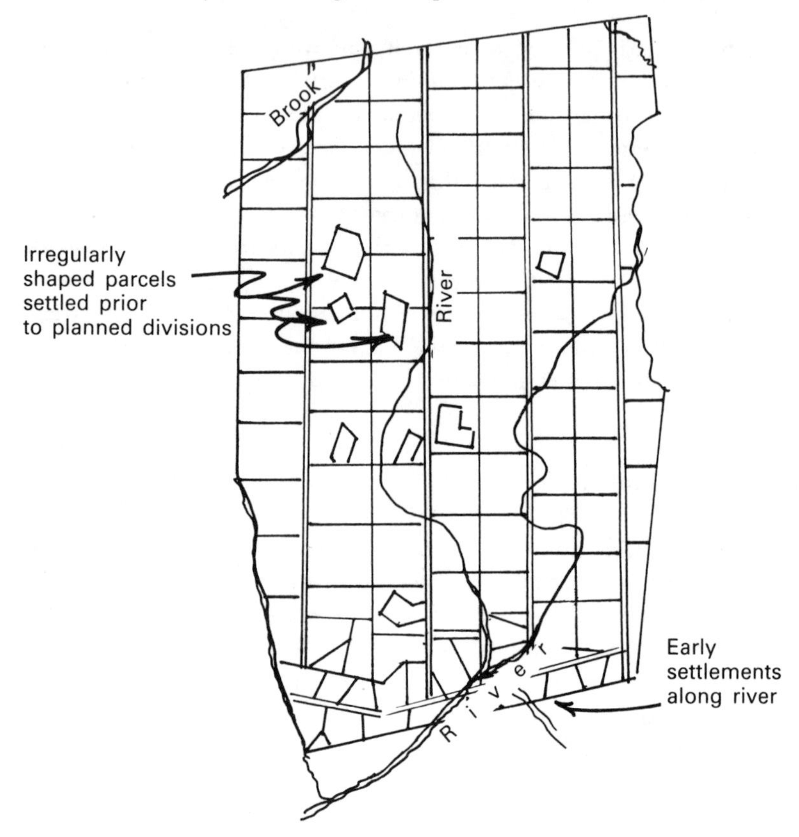

Figure 1 *Rough drawing showing the layout of North Berwick, Maine, illustrating the lower portion of the town as first settled with its irregular shapes, and the upper portion with planned divisions as set forth by the town fathers prior to that section being settled. This layout is typical of many towns in the northeastern United States; however, western states were much more careful to have their towns laid out neatly before being settled.*

the haphazard arrangement of the eastern states. Looking closely at one's town maps will give the same impression. The immediate area of settlement is in general a mishmash of lines running here and there, whereas the outer fringes of towns may show a more orderly plan. When towns did attempt an orderly plan, swamps, trees and large hills had to be circumvented, resulting in roads that meandered and spawned irregular parcels of land.

Seeing this and realizing what was happening, the government instituted plans whereby the western states would be laid out in a more orderly fashion. Meridian and base lines were set up, surveys were made spelling out parcels in distances and bearings, and the plans were put on file.

Deed Descriptions

Early deed descriptions (and some much more recent) were *general* rather than *accurate.* A person selling a parcel of a lot might write the deed as "200 feet by 200 feet by the stone wall." And if dimensions were not known, it was easy to say that the lot ran "by land of Jones; thence by land of Smith," and so on. There was little or no attempt to be specific; professional surveys were not considered that important. After all, "anyone could pull a tape around."

Problems Arising from Population Explosion

With the end of World War II and the financial boom that followed, it came as no surprise that there would be a population boom as well. And a population boom would result in a need for more homes.

Cities were looked upon as places where one worked, and families began setting their sights on those suburban homes that began sprouting up where old farmlands had been previously.

Prices on land began to rise, and developers were searching for and buying up farms for housing.

Planning boards sprang up with the hope of controlling the growth and managing the arrangement of lots. Developers were required to file with the planning boards the plans of their subdivisions for approval. These boards were composed largely of laymen who worked voluntarily after their own long day's work, and it is not surprising that great accuracy was not always demanded. Frequently corners got misplaced. And when bulldozers worked the land, and when poles were installed for phones or posts for fencing, corners that might have been placed were frequently lost. Fences and bush lines went up where occupants thought their lines ran. Those plans that were faithfully drawn up and approved by the planning board and filed with the registry of deeds may or may not have related to the later existing lines.

Inherited Land and Generation Problems

We find that deeds for entire farms were usually written "by land of Jones; thence by land of Smith" or similarly. Later these farms were inherited by the farmers' sons. But while alive the father worked the land and knew the bounds; and whenever he and his son were working the property, he might say when pointing to the back forty, "Now that piece of land is really ours, but I let Joe use it for grazing his cattle." Should discussions come up about the possibility of his placing a corner marker he would, as often as not, say that there would be no problem with Joe because they both "understand that the line of trees with the barbed wire is the property line."

And perhaps the farmer got his land from his grandfather. Yes, he has a deed somewhere, but which one is it among the several deeds he has for this and other land? This becomes a real problem for his widow or son after his death when the property is passed on.

Those are some of the more obvious reasons why deeds may not identify the boundaries and locations of land parcels accurately. There are others, of course. Few states require that a parcel of land being sold must be surveyed; thus a buyer, especially, is leaving himself open to problems if the deed does not accurately describe not only the property but also its location. (Surveyors refer to location as where the property exists "on the face of the earth.") Too often buyers will rush into a sales agreement without first being certain of precisely what they are buying and where. They rationalize that the cost of ascertaining this will be greater than any loss they might suffer at a later date or as a result of a competitive bidder. Experience, however, does not always bear this out.

Purchasing Land

Buyers of land are commonly under the impression when a title search has been completed that there is no question where the bounds of the property are. The problem is that too few people take the time to look at those bounds. Instead, the house is viewed casually, and the water and sewage checked out. If the house and the land have been in use for some time, the bushes along the side are assumed to be the line, and the tree with the barbed wire out back is thought to be the corner.

If the parcel is in a subdivision, we are shown some stakes in the ground or perhaps some iron pipes. We are then shown a little survey plan which has been signed by the local planning board. The deed is usually written up around all of this and we think, "Swell, there is no problem here!"

But now, let's check out a few things:

- Are all corners of the land well marked, and are the boundaries clear as to their exact location on the face of the earth?

- Do each and every one of the abutters involved agree with these boundaries and corners?
- Do your deed and the deeds of your abutters clearly state that these are indeed the corners and the lines?
- Do all deeds refer to specific monuments that can tie your parcel of land to a specific spot on the face of the earth? Could this parcel be precisely located again if the monuments were misplaced?
- If this parcel of land were only a portion of a bigger property, does it meet all requirements of the local and state subdivision laws?

There may also be other points to be checked, and it is likely that a professional surveyor and/or lawyer would be the only people who would know which local laws need to be checked out. The adage "Haste makes waste" can be very true if buyers do not take time to be certain that they are buying what they think they are buying.

Typical Problems

Let's look at some typical problems for the purpose of understanding why a survey is important.

Deed vs. Land. The average deed today has little resemblance to the parcel of land it describes. A small house lot, for instance, may read "100 feet plus or minus to an iron pipe; thence 200 feet from that pipe to another by land of Jones" and so on. If the iron pipes can be found, they seldom check out to the distance mentioned. But more than likely one of the pipes is missing, and a line of bushes has become the accepted property line. Should the parcel be in the woods, the only bounds are likely to be trees with remnants of barbed wire in them, or possibly a stone wall will be thought of as a bound. But the deed will seldom give details as to the specific location of this parcel of land.

Original Owner is Dead. Buying land from the estate of the original owner can be a real problem when no one knows where the exact boundaries are. But the surveyor does contend with, and can help solve, this problem.

Inflated Acreage. Before there were tax maps, aerial photographs, and other methods of determining areas with some degree of accuracy, deeds were usually written by lawyers, with the owner indicating the approximate area being sold. For instance, the owner might pace off the back forty and tell the purchaser he was selling him forty acres; and so the deed would say "forty acres plus or minus." Some land has been found to be "plus" for sales purposes and "minus" for tax purposes.

Other problems result when "plus or minus" property is subsequently divided and sold. It is easy to see how a plot "thought" to be about ten acres could end up being closer to five.

So, a purchase of a twenty-acre parcel, plus or minus, may or may not involve an actual twenty acres.

Why a Survey?

Because of the laxity with which records have been maintained, and because of the indefiniteness with which deeds have been written, a landowner or purchaser of new property runs a risk of losing land. A survey will define the boundaries in precise terms, and it will determine where, on the face of the earth, the property is located so its location can be reconstructed at any time in the future should corner markers and other points of identification be missing. It is not only the property of the owner, or of the potential purchaser, that is likely to have a "more or less" deed, but also that of the abutters, and it behooves the owner to determine what is truly owned before the property is partly taken over by those who adjoin the land.

We shall attempt in the next chapter to give you a glimpse of how the professional surveyor surveys land. Later chapters will develop how he goes about his work and seeks out information to end up with an honest survey.

Chapter 2 *How a Property Survey is Made*

Introduction

The average person probably thinks of a surveyor as the person who, with transit and tape, measures angles and lengths to determine the area of a parcel of land.

Although that is a great oversimplification of the duties of a surveyor, it is one of his or her goals. But in addition to this goal, the surveyor has another goal—locating the parcel in an exact spot on the face of the earth.

In order to understand the basics of the work, let's go into the field with a surveyor and assistants and watch them measuring the angles and bounds and later locating the parcel at a precise spot. We'll assume a very simple parcel of land—a perfect rectangle. Of course this example is greatly oversimplified; far too few parcels are perfect in shape with right angles for corners and with two sets of parallel sides. But this example will give us some background for later chapters when we shall be concerned with the real world of the surveyor, where there is a question about the accuracy of the deed, where the sides may curve and the angles vary, where hills and brooks present special problems.

Let's assume then that the surveyor has a deed describing the parcel as follows:

> Beginning at an iron pipe on the northerly side of Main Street, Town of Jersey, County of York, State of Rhode Island; thence running along said Main Street S 45° E for a distance of 80.0 feet to an iron pipe at the southwest corner of land now or formerly of Jones; thence turning and running N 45° E for a distance of 100.0

> feet along land of said Jones to an iron pipe; thence turning and running N 45° W for a distance of 80.0 feet along land of Smith to an iron pipe; thence turning and running S 45° W for a distance of 100.0 feet along land of Kelly to an iron pipe which is the point of beginning. Said land to contain 8,000 square feet, and all bearings refer to magnetic north of 1954.

You can see that this description would be drawn as shown in Figure 2.

Field Survey

We shall assume further that the parcel contains no obstructions to hinder sight or the taking of measurements. We will use the

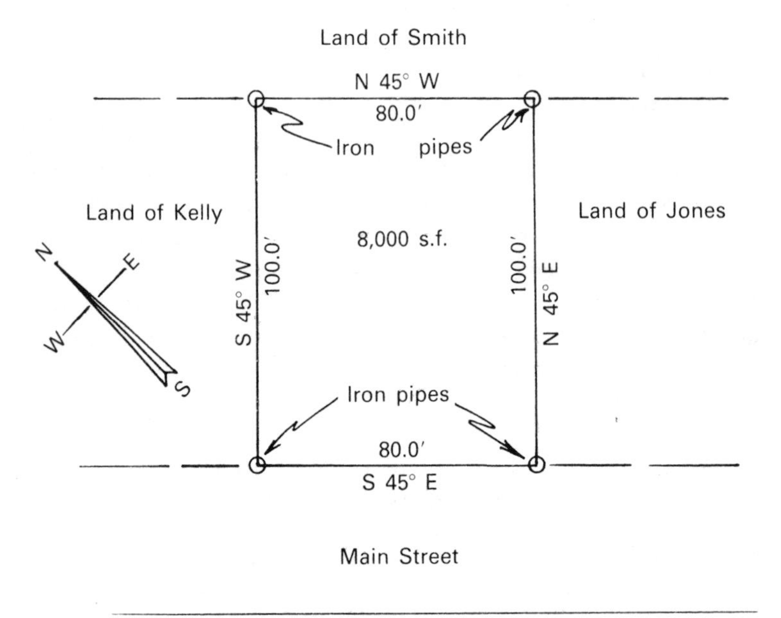

Figure 2 *Showing parcel described in deed description.*

standard transit and steel tape and, since we have no obstructions to contend with, we will use a small hand target to place on the corners for sighting in accurately with the transit. Most of us are familiar with the transit, having seen them used on roads, bridges, and buildings. The transit fulfills the need of a sighting instrument, with a telescope added for long-distance sighting (see figure 37A). This revolves on a flat metal disc divided into 360 degrees, and each degree is divided into 60 minutes. Minutes can further be divided into seconds by a graduated scale called a vernier. We have all used a simple protractor in school which enabled us to lay out portions of a circle; this is exactly what we are doing here as we sight the scope on one target and turn to another to get a reading on how many degrees we turned. Many modern transits, however, provide digital readings (like a modern watch in degrees, minutes and seconds.

Point of Beginning

Commencing at the front left corner pipe, or the point of beginning as noted in the deed, the surveyor sets the transit up over this point. A plumb bob, suspended by string from the head of the transit, is zeroed in over the exact center point of the corner pipe. Leveling and centering over the corner as precisely as possible is important. Without this accuracy, an error can be brought into the angle shots and will later show up in the total calculations of the work.

Now the surveyor sights in the cross hairs of the scope over to a target that the rodman holds on the back left corner. By setting zero on this point and then swinging the scope from that corner to the front right-hand corner, he gets a reading of 90°, which checks with the deed's bearings.

To perform a more precise survey, the surveyor can then lock this angle on the upper plate of the transit, then release the bottom plate and return to the first point. After zeroing in on this point, the top plate is released and the angle turned again. This angle should now be double the first. For example, if the first was 90°, 45′, 20″ (ninety degrees, forty-five minutes, and 20 seconds), the second angle should read 181°, 30′, 40″. The surveyor can repeat this process as many times as desired and divide the number of degrees by the number of turns to verify that the angle he got on his first turn is correct. Using an instrument which gives digital readings, rather than a disc and vernier scale, makes these checks faster and easier to perform.

Figure 3 shows the transit set up over the front left corner or point of beginning. A sight is taken on the back left corner pipe, and then the scope is swung to take a sighting on the front right-hand corner pipe. The intent here is to get an accurate interior angle reading, so the surveyor can sight the front right corner first and then swing the scope to the back left corner and obtain the same result.

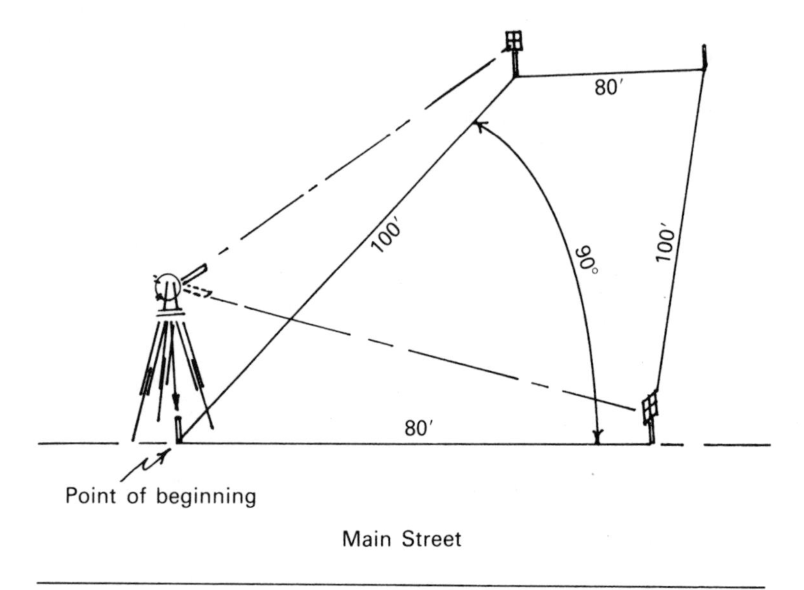

Figure 3 *Transit in position to determine angles to corner markers.*

While still maintaining this position with the transit, he takes compass readings on both the left-hand line and the front line. The bearings in this case would be S 45° W for the left-hand line and S 45° E for the front line, providing there are no distractions to the compass. This then gives a check with the interior angle of 90°. See Figure 4.

Measurements are also taken of these two lines and are recorded in the field book.

Now the surveyor moves to the front right corner with the transit and sets it up over the pipe at the previous corner, sighting back to the front left corner and turning the instrument to sight the right rear corner.

Bearings and dimensions can now be taken of this right side line and noted in the field book. The interior angle would of course be 90°, with the line being 100.0 feet long. Proceeding to the next

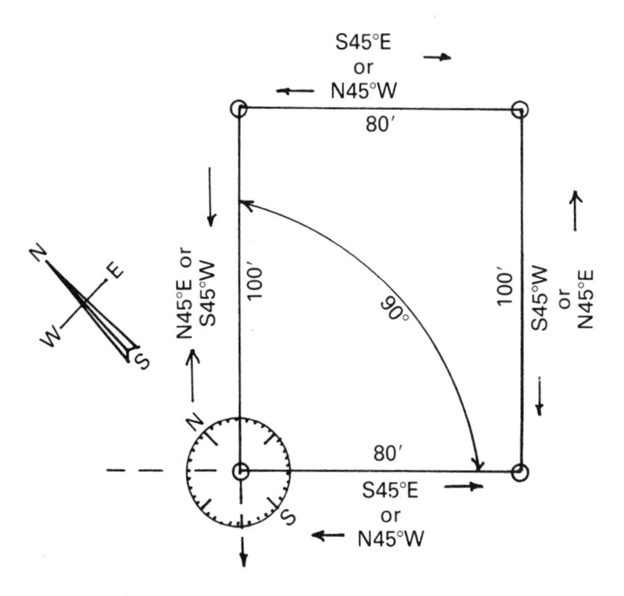

Figure 4 *Surveyor now takes a bearing of the line by determining the angular value on the graduated scale of the compass.*

corner (the right rear corner), he then obtains this interior angle and the back bearings and dimensions of the lines, and he next takes a final reading over the back left corner for an interior angle reading.

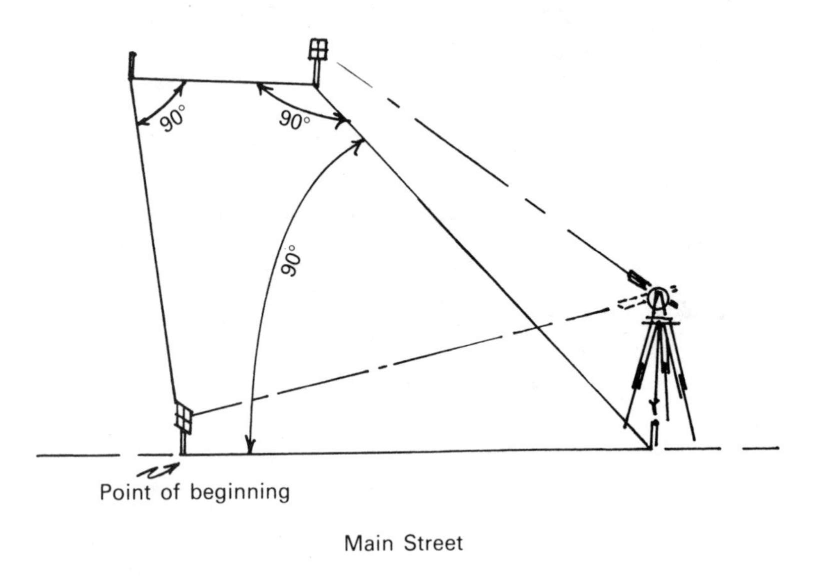

Figure 5 *Surveyor continues to other corners.*

Field Notes

The field survey notes and drawing would therefore look something like Figure 6.

It contains the name of the owner of the property and its location. The date, weather conditions and the initials of the transit man, or so-called party chief, are noted, along with the initials of the rodman and any other helpers on the job. Weather data are noted, because today's more precise instruments must be

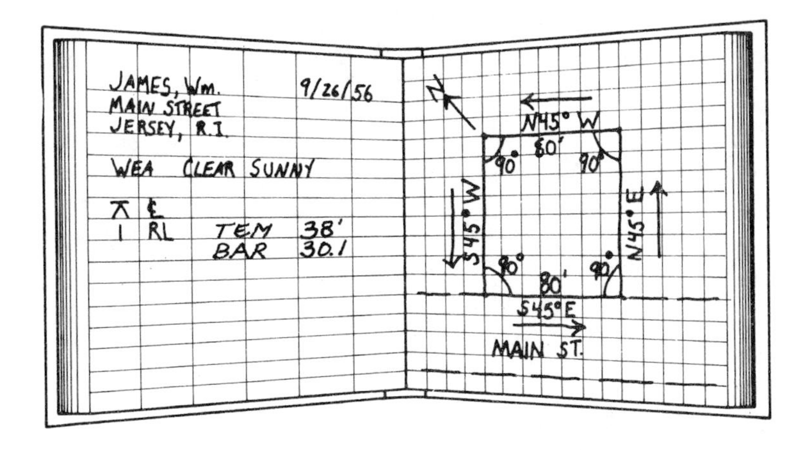

Figure 6 *Simplified illustration of field survey notebook.*

adjusted for the temperature and barometric pressure at the time of the survey. It is important to note the date of the survey, since changes over time in the location of magnetic north can invalidate instrument readings—even a few years can make a difference. If future surveyors know the date of a survey, it will be easier for them to convert readings and bring it up to date. Any special conditions or problems that may have arisen on the job would also be noted.

Realism

As we all know, there are very few boundaries that are as clear-cut as those in this example, and the typical deed is not spelled out with as much description, so let's see what happens when a parcel of land is bounded on the front by a road, on the right and rear by a stone wall, and on the left by some trees with some barbed wire in them. Assume that this wire is strung fairly straight from that back stone wall to the road. This might look as shown in Figure 7.

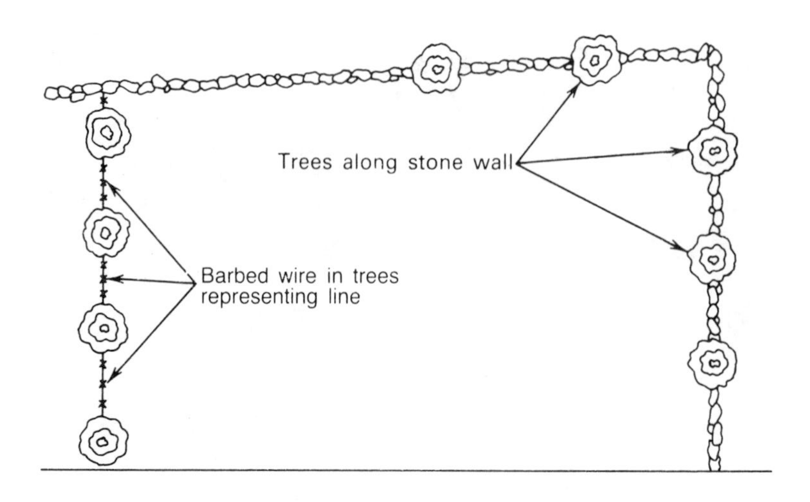

Figure 7 *Illustration of typical job with trees in line.*

With the stone wall along the right and rear of this parcel, we probably will find no pipes in the ground for corners. The left side has those trees, and so a tree at the back left and at the front right with the wire in them have most likely been assumed to be the corners. The first step would be to get all abutting owners to agree on exact points where pipes could be driven in for corners. We know that the intersection of that stone wall is the corner, but unless it is a large stone we believe will not be moved, there is no way of marking or picking an exact spot to run the line. The same applies to that tree at the road, which would best be changed to a pipe driven alongside, in front or in back of it, depending upon where all abutters agree.

Because of the stone walls and the line of trees, it is not possible to run a tape measure accurately along the bounds of these three sides. Nor is it possible to sight a transit through a line of trees. So we resort to a method known as triangulation.

When you studied trigonometry, you learned that if you know the dimensions of two sides of a triangle and the included angle, you can determine the length of the third side. In addition, you learned that if you know two angles of a triangle and its included side (length), you can determine the lengths of the other two sides. You can also determine the third angle by deducting the sum of the two known angles from 180°.

With these principles of trigonometry in mind, it is simple for the surveyor to set up three points inside the parcel (call the points B, D, and F) and draw imaginary triangles that either have two known angles and an included known side or that have two known

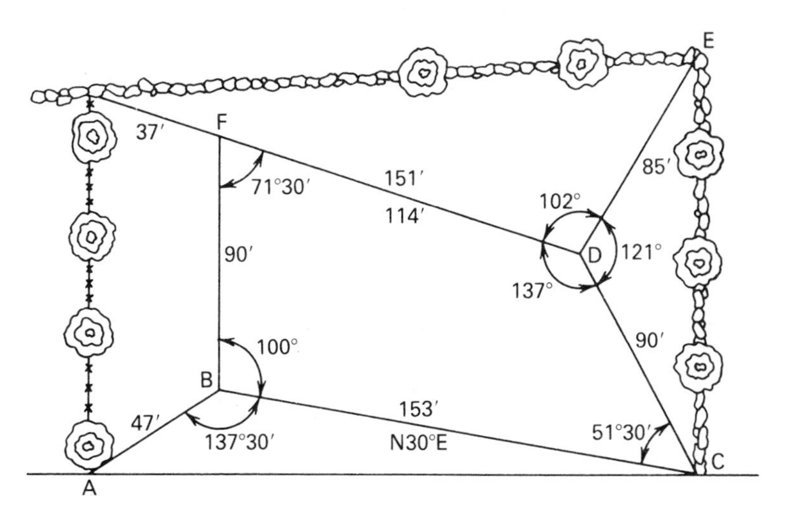

Figure 8 *Illustration showing method of running offset lines. With instrument set at points B, D, & F, surveyor can obtain data needed to compute perimeter lines of property.*

sides and a known included angle. With this information, the dimensions of the stone walls and the line of trees can be figured to fractions of an inch. Present-day calculators help immensely.

Lines for these triangles are called offsets. Arrangements of these would vary in respect to the site conditions, but Figure 8 will give the general idea.

Standard Surveys—Problems

The stone walls and the line of trees mentioned in the above example are only two examples of obstacles the surveyor frequently finds in the line of sight. There are often buildings and other man-made obstacles, and knowledge of triangulation helps in measuring "through" these obstacles.

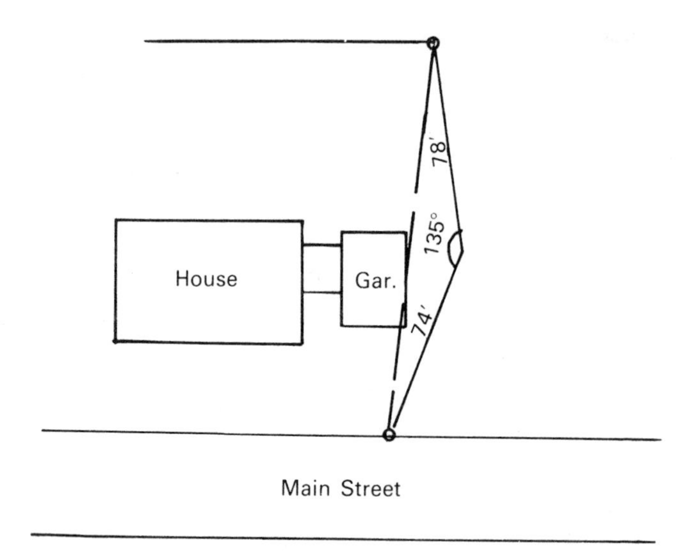

Figure 9 *Illustration showing method of running line when building is in the way. Here we made use of the formula based on two sides and an included angle.*

Another problem is represented when there are hills and vales in the surveyor's line of sight. Such sightings must then be taken in steps as shown in Figure 10.

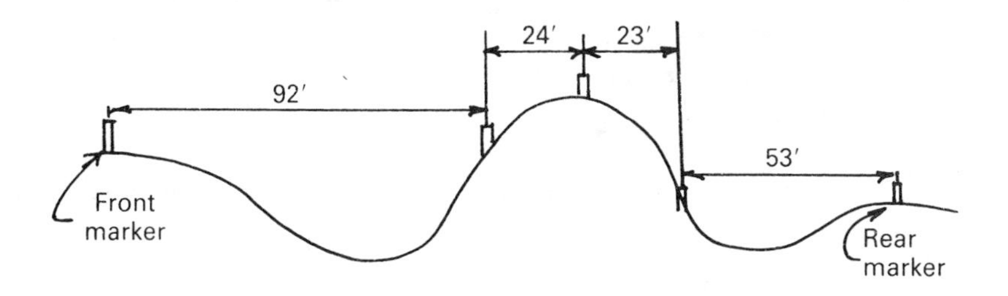

Figure 10 *Survey points along line showing horizontal measurements.*

Horizontal Measurements

Dimensions of land must always be based on horizontal measurements. Let's take a look at what would happen were we to stretch our tapes up hill and down dale and then add it up.

If you add up all of these dimensions, you'll find that they come to 337 feet, which is a far cry from the 266 feet of horizontal dimension.

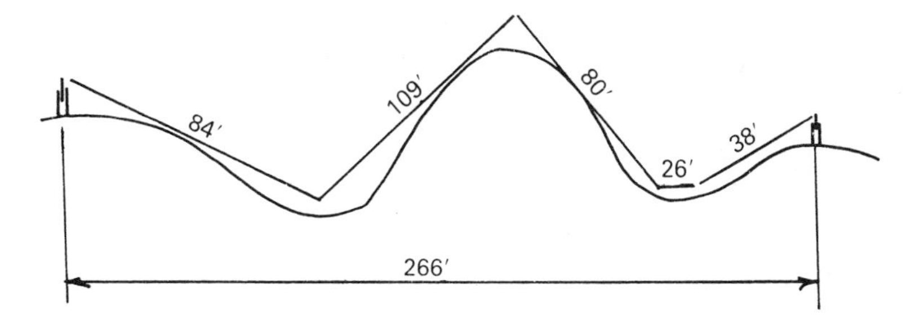

Figure 11 *Measurements over hills and valleys distort the true measurement.*

Location

A major point which has been a problem even in some of the finest surveys performed is the tie of land to the face of the earth. Note in the description and in the plan layout of the first parcel discussed that all of the corner pipes were in. We were told where the parcel existed in regard to the state, the town, and even the street. But, as you know, some of our main streets are rather long. Consideration must therefore be given to some means whereby this parcel of land can be tied down with reference to something permanent such as a state highway marker.

But now what about the job where you haven't any pipes to refer to? Or how about that stone wall and barbed wire in trees when the people next door do not agree on that line of wire in the trees? Or perhaps even that stone wall?

Research

When abutters do not agree on those lines, or when there just are no markers to go by, a surveyor is sorely needed. And so, handed a deed of vague description and told only approximately where corners should be, the surveyor is requested to provide answers that he must be prepared to support in court, if need be! Our next chapters will tell you how he proceeds.

Chapter 3 *Aerial Photos— Tax Maps*

Puzzles

We are all familiar with jigsaw puzzles. We search through the pile for pieces that fit into one another. Perhaps we start by completing the outside edges and filling in the picture from there, while always looking for a certain curl or shape that will mesh with another piece, until we finally complete the picture.

Try to visualize this on a larger scale. Take all the deeds that describe parcels of land in a particular township and attempt to put this puzzle together into something shaped like the town. With hundreds of landowners working, each with a piece that he believes is the shape of his property, it becomes a puzzle in itself just trying to fit together all the deeds. Imagine, too, the problems that develop when the deeds read "by land of Jones; thence by land of Smith." And there would be further complications if the deeds read "more or less" for acreage.

But this puzzle is not an exaggeration. Towns have just not had much luck defining all parcel borders within their confines until recently.

Town Mapping

World War II changed this, however. The technological improvements in photography allowed reconnaissance planes to photograph vast areas of terrain so enemy locations could be spotted. The detail from these pictures was further refined by more sophisticated developments in photography until today such photos indicate not only rivers, forests, and open lands, but individual streets, houses, and sidewalks. Fences, too, show up when not hidden by trees. Even variances of tree growths can be seen from these photographs, as well as land under cultivation and pasture land. And still greater detail is possible.

These photographs, because of their great detail, are actually aerial "maps" which cannot be matched by individual sketches. So at last physical maps of a township are possible. This is a godsend to assessors, who need to be assured that *every piece of land within a township is taxed.* Accordingly, these maps have come to be called "tax maps."

Today a surveyor, too, makes great use of aerial maps. In fact, they are among the first things he wants to examine after being asked to do a survey. With deeds in hand and blown-up photos of a specific area, he can validate the boundary lines. Of course he must also consult rights of way and other surveys that may be on file, and do an on-site field review.

The Cost of Accuracy

Aerial maps are not inexpensive. If commissioned to be done, $40–50,000 (depends upon size and congestion of populated areas involved) might not be an out-of-line cost for a particular township. This could then be the first time a township ever had an accurate method for taxation. There may never have been a way to validate the deeds theretofore, and when we remember that most of them have been written in the same old way down through generation after generation, we may conclude that they were nothing more than a bunch of pieces of a puzzle uncoordinated with one another.

So the cost of the maps to a town must be weighed against what additional tax revenue the town may enjoy in future years as a result. Also, the actual cost will vary, depending on the amount of detail that is desired.

Naturally, if these parcels of land are in a wealthy town, more could be spent on the aerial survey than we would want to spend on one in the north woods.

Nevertheless, good aerial maps will allow a township to lay out all known parcels of land on it; then the others should be investigated for ownership and potential tax income from the rightful owner, since the town, even though it cannot account for each and every parcel of land by a deed, is nevertheless taxing some taxpayers for it.

Many areas already have aerial photos taken by the government, which are available through the U.S. Department of Agriculture. These can be obtained at a minimal cost. Although they may not be practicable for town "tax maps," they are handy for use in the surveyor's business.

Public Viewing

The common way to use aerial maps is to work from the known to the unknown. Known parcels (those with firm deeds) are outlined on the map. The "unclaimed" parcels must be identified, and a researcher will seek out all available data such as unfiled deeds, sketches, field survey drawings, etc., that are not on file and that owners and abutters may have. The maps are then laid out with all known data on them.

Any remaining land must be further researched for ownership. At this point the procedure is to advertise and notify by public announcement that these maps will be spread before the public for review. They are asked to come in, check their areas for any possible errors they can prove or disprove, and to suggest the ownership of "unclaimed" property.

The aerial maps then become the basis for taxing the town's property owners.

With this as a start and with corrections made whenever surveys are conducted thereafter, eventually the town knows each and every parcel outline and the owner thereof.

Not an Original Idea

A good town manager would probably make a case for accurate land surveys somewhat as follows:

> "Half measures always result in a loss of time and money. The only way to sort out the confusion in the field of general land records is to proceed with the surveying and evaluation of each individual land parcel in all the communities of the Empire. A good cadastre [technically means "notebook" but refers here to systematic surveys] will constitute a complement of my Code as far as land possession is concerned. The maps must be sufficiently precise and complete so that they could determine the boundaries between individual properties and prevent litigation."
>
> —Napoleon Bonaparte, 1807

Other Uses

There are of course many other uses for aerial land maps which should also be considered as offsets to cost.

State Highway Departments. Such maps are indispensable for projecting new roads and superhighways. Immediately it becomes obvious if a new road will disrupt a town up ahead, or if a highway can be bent to avoid the cost of cutting through high hills.

Town Highway Departments. They, too, can use these maps for straightening roads; to remove a curve may be more or less expensive than taking a longer sweep around an obstacle. Or would the town be better off to do only a few sections at a time?

The usefulness of aerial photos is illustrated by the experience of a nearby town. Town officials decided to complete the middle section of a partially constructed road. Initially, they intended to continue building straight out from one finished section. Examination of an aerial photo, however, revealed that if they did so, they would eventually have to make a sharp turn to link up with the other completed section. Thanks to the aerial photo, town officials were able to plan the overall road and then work on short segments as funds became available.

Fire and Ambulance Departments. Fire departments find these maps of great value when plotting routes to reach fires. So do ambulances when moments are precious in saving lives.

Official Town Boards. Features such as location of a brook, planning future road systems, zoning, subdivisions, boundaries of flood plains, fire lanes, snow removal routes, bus routes for school children, and locations of utility service lines are uppermost in consideration for a well-planned township. Planning boards, zoning boards, and town and school officials can do much more accurate and efficient planning when aerial maps are available.

Real Estate Brokers. Real estate brokers see them as a very useful tool to describe the lay of the land, road systems, etc. to a client.

Land Surveyors. A properly keyed map can be of great help to indicate to the surveyor the location, abutters and general layout of the parcel to be surveyed.

Geographic Information System (GIS). Aerial maps provide the base for the new Geographic Information System (GIS), increasingly being used by states, counties, municipalities, and planning committees. With GIS, any number of layers showing utility lines, roads, soil types, etc., can be superimposed on the base to provide whatever combination of information a community or committee needs. (See Chapter 12 for more information on GIS.)

Public Responsibility

In spite of the value of land maps to so many, they are no better than the degree of responsibility of the public to maintain their

Figure 12 *Typical aerial photo. Note how outlines of fields and woodlands can be discerned. Roads are detectable, and in the lower right-hand corner a railroad line cuts across the photo on a 45-degree angle.*

accuracy. Any past surveys that our forefathers may have had done but were not filed in the registry should be brought forth. Any new surveys should be reported. There are cases where a landowner is paying taxes for, let's say, fifty acres when he knows his neighbor is quite innocently paying taxes on his other fifty.

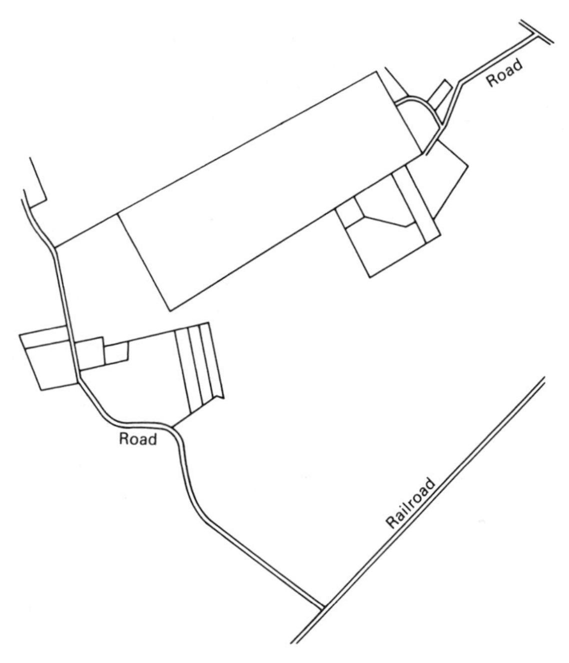

Figure 13 *Assumed lot lines taken from the aerial photo.*

Example

Figure 12 shows a typical aerial photo and Figure 13 shows how property lines are developed from it. These are the obvious lines due to tree lines, stone walls, etc. They are then corrected by public viewing. Perhaps one farmer owns three or four of those fields or maybe just one stone-wall-enclosed parcel.

Examples of road curves and road layouts that should be considered by planning boards when approving developments back-to-back or just on winding roads are shown. Will that curve come out in road plans in the near future? Should the developer be allowed to build the housing near it? Would subdivisions be back-to-back and thus offer savings if roads could be designed to interconnect for school bus routes, snowplowing, etc.? Aerial photos can help town officials determine whether or not to work with adjacent towns to lay out roads so that they will link subdivisions in a logical, economic way.

Chapter 4 *Researching a Parcel of Land*

Assembling Data

Researching a parcel of land normally commences when the initial contact is made between the land surveyor and client. Tax maps, if available, are a good starting point. The client or the surveyor makes a tracing of the area from the tax map, and any notations, deed markers, names of abutters, and any other pertinent data are noted. When tax maps are not available, the client's sketch has to suffice.

Data gleaned from the tax map or from the sketch will indicate information that should be confirmed as well as information that is missing and should be discovered.

Walk of Lines

Surveyor and client should then "walk the lines" (boundaries) of the parcel as described, giving the surveyor the client's "interpretation" of the parcel itself. Any corner monuments that are in place can be flagged with ribbon, and any missing corners can be temporarily flagged; the surveyor's notebook should record these. Other notations, such as boulders, barbed wire fences, trees of large size in a line with smaller growth on each side (indicating a line, especially if some barbed wire can be found protruding from the trees) are also recorded in the surveyor's notebook.

Abutters

Even when there is no argument with abutters about boundaries, it is always good business to inform them that a survey is being undertaken and invite them to walk the lines with the surveyor and client. When there is a boundary dispute, it is even more important that this be done. After all, the accurate boundary lines are also the lines of the abutters. Furthermore, contact with them can prove very fruitful. Sometimes they will pull forth deeds which have never been filed at the registry, and they may have survey plans of their own parcels that were not filed which can provide information that will save time and money in research.

Almost invariably abutters are as happy to settle the line questions as the client. As for expense, it costs them nothing, since the client is paying for it.

No matter how much data there may be concerning a parcel, and no matter how enforceable these data may seem to be as far as pinpointing bounds, if an abutter wants to dispute a line he may do so. This will, of course, lead to both parties' engaging lawyers and engaging in litigation. Because of the costs this could entail, it is much wiser to settle disputes on the spot if possible. When deeds are of little help in describing the bounds and a dispute arises, if a settlement between the client and the abutter can be decided on the spot, the surveyor will note all agreements in the survey. The plan will then be signed by both parties prior to filing. The client and the abutter should follow up by having their deeds rewritten to conform to the agreement.

Generations Dying Off

The earliest settlers had no way of registering deeds or even writing a deed down. So they walked their lines with their sons, pointing out identification marks that were more or less permanent. And these sons walked the lines with their sons and passed the information on from generation to generation.

The same method has been used to some degree by the fathers and grandfathers of many who are living today. They walk the lines with their offspring or younger relatives, pointing out identification marks and what they feel are their rightful corners. At the same time they may be heard to say that they plan to have the land surveyed "some day."

But many of these fathers and grandfathers are in their seventies and eighties, and that "someday" never comes. They represent a generation that inherited land in the early 1900's. Suddenly a widow is left standing with a handful of deeds and with little or no knowledge of what they all mean or where the land is located.

That is one generation we are losing, but there is another equally valuable generation we are losing—those men who cut wood on land at the time *individual parcels* were being lumbered—not the widespread operations of log cutters today, who wipe out all areas in one big swath. Those loggers knew the boundaries of the individual parcels they worked on and who abutted whom. They knew which Smith, or which Jones, abutted whom and where his current descendants lived. They knew that a particular large oak tree at the brook was a corner, or that the boulder at the base of the ledge was a common corner with the abutter. As a research source, those loggers were invaluable, and they still are when they can be found. But they are also dying out.

Lawyers

Frequently lawyers find flaws in a title and request a survey. Any research a lawyer may have had done should be obtained and studied. It may save the client's time and money since duplication can be avoided.

Registry of Deeds

Once the surveyor has the sketch as well as copies of any deeds, file numbers of deeds, and any other related data found during the field inspection and conversations with abutters and others, he then goes to the registry of deeds to make a personal study of all deeds relating to the parcel, and this includes deeds relating to adjacent parcels. Not only must he interpret the various deeds as they pertain to the parcel under investigation, but he must also assure himself that he has located all pertinent data relating to the abutting parcels. Information regarding corners and bounds can be added to the sketch the client provided.

Should the surveyor ever be called into court to defend a survey, his testimony will be no better than his efforts to search out all information relating to the parcel.

What Is the Registry?

The main function of a registry of deeds is to provide a place where all transactions regarding land sales, liens, mortgages, and attachments to a parcel of land are recorded. It is managed by the registrar of deeds, who records all such transactions, and it is open to the public for inspection and use. Should the original copy of a deed become lost, the registrar can certify that the copy on file is a true copy of the original deed.

Types of registries and their locations vary from state to state. Some registries are in individual towns; in others they may be in county offices; in still others they may be in state offices. The conditions of registries vary from poor to excellent. One may have only a pile of books in the corner of a back room where entries are made; still others may have complete filing systems and modern machines to reproduce deeds.

If the registry is in a town office, there is likelihood that the town clerk is also the registrar. In county offices, there may be one person in charge of the registry, and in state offices the registrar may even obtain the position via the ballot box.

Because of so many variations from state to state, only your local registry can advise you what information, if any, you are required by law to file there.

Recording vs. Registering

There is a difference between a *recorded* deed and a *registered* deed.

In general a person can walk into a registry of deeds and *record* anything at any time. For instance, you could make a rough sketch of your property with all the information you know and have it filed and recorded, but it would not be official. Many towns now forbid a landowner to subdivide property and record it himself; engineering and surveying stamps are required.

To be *registered,* a deed must have been completely cleared of all possible questions regarding property lines. A complete survey conducted by state-approved surveyors must first be performed, with all lines being definitely established. Then this survey with supporting data must be carried through court proceedings in order to establish a complete legal approval of property lines.

Inasmuch as few states have the facilities for *registering* surveys, most surveys are *recorded* only. For this reason, it is important that when a parcel is surveyed, the lines are approved also by the abutters. This obviates future problems should a lost deed suddenly show up.

Typical Filing Method

Most registries have a set of books with entries running in sequence from book number one to the current book number. Within each book, the deeds are recorded chronologically as they are received, starting on page one and continuing until the book is filled. The early books were of a heavy, bulky type, with the deeds handwritten in ink; today we are most likely to find a small book with typewritten pages.

If we do not know the book in which the deeds we want to examine are located, how can we find them? Most registries have what are called "Grantors" and "Grantees" index books. If the parcel we are concerned with was sold by Jones to Smith, we can look in the grantors (sellers) alphabetical list for the sale by Jones at the approximate date we think the sale was made. Or we can look at the grantees (buyers) alphabetical list similarly for a purchase by Smith. At either of these listings we'll find references to the registry book—page number, date of sale, and date the deed was recorded. Now we can go to the registry book itself and examine the deed that was filed.

If there are no grantors/grantees indexes, we would have to search through the registry books themselves, looking for the approximate dates we have in mind until we find the deed.

Recorded Plans

Since a registry of deeds is a recording facility for matters pertaining to land, it will record and file layouts of land. These are filed in what is known as the "plan books." These plans, whether done by a registered land surveyor or the owner of the land, are usually required to be drawn on linen or Mylar and in ink. The fee one pays for recording usually depends on the physical size of the plan.

When the surveyor files this plan, he also takes along additional prints of the layout (authentic copies) which are signed by the registrar. The registrar notes on these prints the time and date received as well as the book and page number within which the plan is filed in the registry. The original linen drawing (which nowadays is usually prepared on a material called mylar) is then placed in the registry files with a copy of same set out for public use.

Figure 14 *Typical Registry Layout.*

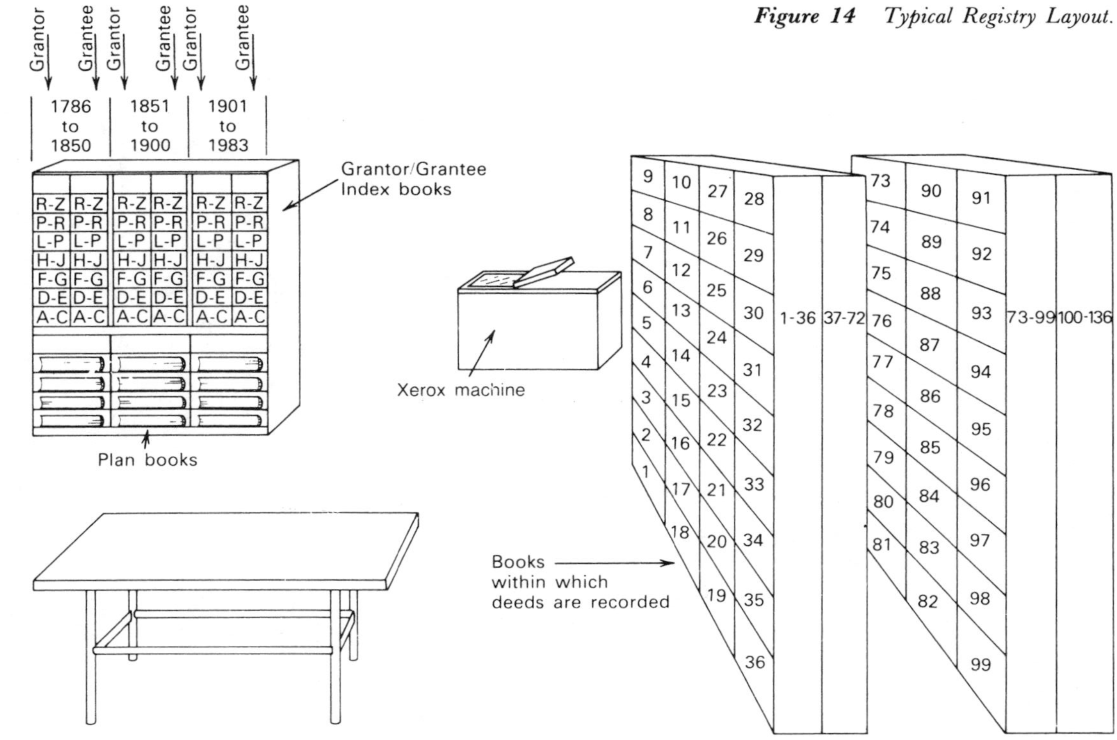

Figure 15 *Typical Grantee Book.*

Grantee	Grantor	Town	Type*	Book	Page	Date	Street**
Jackson, Charles D.	Harrigan, Charles E	Smithtown	Wty	1788	23	Oct. 6, 1879	Thames
" ", David L.	Orville, John S.	Jonestown	QC	1687	324	Dec. 8, 1888	Duke
" ", Edward C.	James, X.	Lakeville	Mo	1879	400	Jan. 7, 1900	First
Jones, Tom T.	Smith, John E.	etc....					

By noting the book and page number of the registry book, you will find the deed which you are researching. If the deed has been written properly you will find references to any survey plans that may have been recorded.

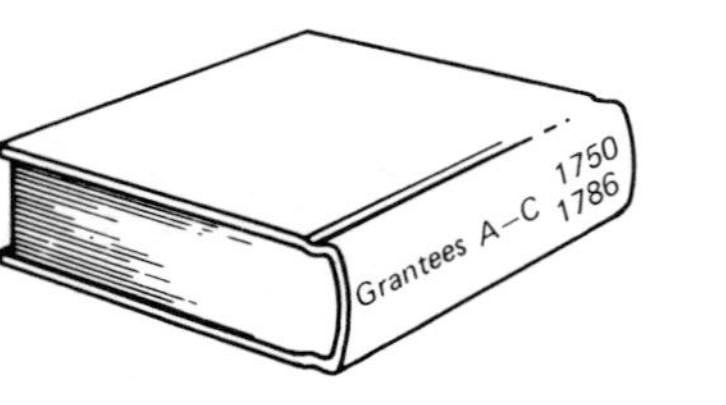

*
Wty — Warranty deed
QC — Quit-claim deed
Mo — Mortgage

** — Street names refer to the street upon which the parcel of land is located.

Any deeds written by a lawyer should now refer to the plan in the description, and this plan becomes, basically, part of the deed.

It is also a good idea to give copies of the plan to all abutters. They can keep it with their deeds, and when their land is surveyed, they can refer to that plan. Giving plans to abutters helps ensure that the jigsaw puzzle of different landowners' property surveys eventually fits together.

Recorded plans are also of great help to any surveyor who is researching records, inasmuch as they may indicate data on dimensions and bearings, or perhaps nothing more than the mere fact that a property line is a stone wall. If the survey plan is being filed merely to update the present deed and no new deed is going to be written, it is well if a note is made in the column of the book and page number where the original deed is written in, stating that there is now a survey plan on file in relation to this parcel of land.

In today's world, when so many problems are showing up regarding land boundaries, banks often ask the surveyor to file a linen with the registry of deeds from which the lawyer can write a better description. They then have better knowledge of what they have loaned money on.

Dimensions from corners of the land to house foundations, from corner points to state highway markers, or to any fixed markers can be helpful to anyone who years later has to replace a lost corner marker. All of this can be noted on a plan but not written easily into a deed.

Confusion

Often, when a client asks that a survey be done so a lawyer can write a deed for sale or purchase, he fails to request that a linen be filed in the registry.

Subsequently the client may discover that the attorney has written a deed referring to a "plan on file in the registry." The only options at that point (since the standard prints the client received cannot be filed) are to have a linen drawn up and placed on file or to have the deed rewritten to eliminate the reference. Either course is an added expense to the client. Most surveyors today routinely file a plan, but it is best to confirm this with your surveyor.

Ready to Survey

With the background information from study of deeds, sketches, plans, and information from abutters and knowledgeable individuals, the surveyor is now ready to return to the parcel of land and meticulously determine how it is really shaped.

Chapter **5** *Mathematical Closure*

Background

Movement on the surface of the earth is always either north, south, east or west, or in a direction in between, such as northeast, northwest, southeast, southwest. It is impossible to move without going in some direction that can be indicated by a compass. For this reason, when the surveyor describes the boundaries of a parcel of land, he not only measures the lengths of the boundaries but he also indicates the direction of the boundaries in terms of north, south, east, and west. Figure 16 is a simple parcel of land. It will show how this is done.

Let's assume it is a square with sides 100 feet in length; since it is a square, all angles are 90-degree angles. The arrow will orient the parcel in terms of magnetic north. The corners are lettered A,B,C, and D for purposes of identification.

In terms of north and east, we'll assume that the line AB is N 45° E, which means the boundary goes north at the same time it goes east but at a 45-degree angle from magnetic north. Similarly, line BC may be described as N 45° W, line CD is S 45° W and line DA is S 45° E. If these were not right angles (90-degree angles), the number of degrees would most likely be subdivided further into 60 minutes (which equal one degree) and into 60 seconds (which equal one minute).

There is another interesting thing about this parcel of land. As we shall see when it is measured, the distance one goes north is the same distance one goes south to arrive back at the starting point, A. And when one travels eastward, one travels the same distance in that direction as he will need to go westward to arrive back at that starting point, A. It's much like throwing a ball into the air; it

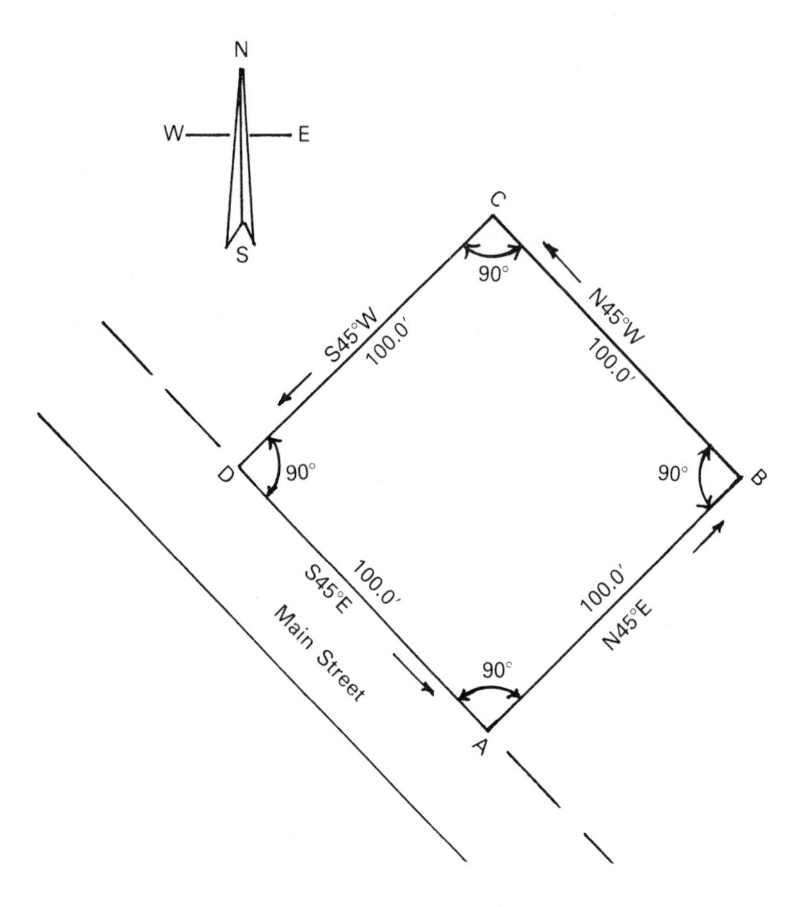

Figure 16 *A simple parcel of land.*

travels downward the same distance that it went up. This is true of any parcel of land regardless of its size and shape; the distances north and south are equal, and the distances east and west are equal. This is a very important principle of surveying that must be kept in mind. It is what surveyors refer to as "mathematical closure," for if the distances were not equal, then at least one of the sides would not meet an adjacent one at the starting point.

Now take a look at the irregular parcel shown in Figure 17, noting its many angles and several sides. If all the north measurements are added up, they will equal all the south measurements. So, too, will the sum of all eastward measurements equal the sum of all the westward measurements.

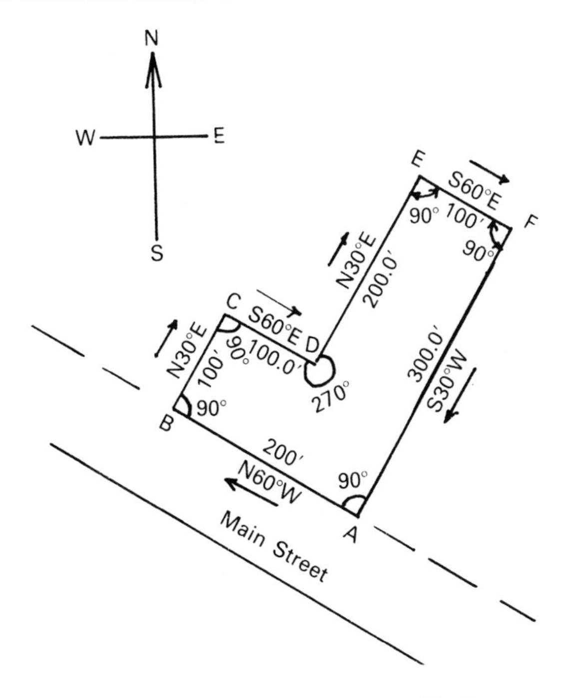

Figure 17 *An irregular parcel of land.*

Interior Angle Test

Surveyors have a field test to determine if a parcel has been "closed" completely with angles. A formula is used to prove what the sum of all interior angles should be. Using the first illustration, we know the sum of the four 90-degree angles will be 360 degrees. But what should the sum of all the angles in the second illustration be? The formula is $180(n - 2)$ where n equals the number of sides in the parcel. Since there are six sides in the second illustration, there should be 720 degrees to be accounted for as follows:

$$180(n - 2) = (6 - 2) \times 180$$
$$180 \times 4 = 720°$$

And we note from our drawing, Figure 17, that 720 degrees is what our surveyor had.

Office Computations

And so, now knowing the bearings, dimensions of each line, and the interior angles of each corner involved, the surveyor can return to the office to compute the actual "closure" of the land. To do so, he makes use of sines and cosines as found in trigonometry. See note following Figure 19.

Figure 18 will illustrate the computation table for our four-sided parcel, and Figure 19 represents the computations required for the six-sided parcel.

Here, as in Figure 18, the north equals the south and the east equals the west total; thus it "closes" or "balances."

Surveying in Real Life

The illustrations we have used are accurate examples of mathematical closure. They illustrate complete closures of simple and compound boundaries.

However, when a surveyor goes into the field to survey a parcel, things frequently don't work out so exactly as in the illustrations we have just used.

Our surveyor probably used a transit and tape—or a modern version of the transit called a theodolite, equipped with an electric

Line	*Bearing*	*Distance*	*Cosine* *Sine*	*North*	*South*	*East*	*West*
A-B	N 45°E	100′	.70711 .70711	70.711		70.711	
B-C	N 45°E	100′	.70711 .70711	70.711			70.711
C-D	S 45°E	100′	.70711 .70711		70.711		70.711
D-A	S 45°E	100′	.70711 .70711		70.711	70.711	
				141.422	141.422	141.422	141.422

North 141.422
Minus South 141.422
0.000

East 141.422
Minus West 141.422
0.000

Figure 18 *Computation for a Four-Sided Parcel.*

Line	*Bearing*	*Distance*	*Cosine* *Sine*	*North*	*South*	*East*	*West*
A-B	N 60°W	200′	.500000 .866025	100.000			173.205
B-C	N 30°E	100′	.866025 .500000	86.602		50.000	
C-D	S 60°E	100′	.500000 .866025		50.000	86.6025	
D-E	N 30°E	200′	.866025 .500000	173.205		100.0000	
E-F	S 60°E	100′	.500000 .866025		50.000	86.6025	
F-A	S 30°W	300′	.866025 .500000		259.808		150.000
				359.807	359.808	323.205	323.205

Note: The sine is a trigonometric function pertaining to the ratio of the side opposite a right triangle and the hypotenuse of the triangle. The cosine represents the relationship between the side adjacent to a given acute angle in a right triangle and its hypotenuse. Sine and cosine equivalents for various angles are available in published forms.

Figure 19 *Computation for a Six-Sided Parcel.*

distance measurer (EDM)—instead of the older compass-and-chain method for the measurements, but even so, minute inaccuracies can occur to spoil a "perfect closure," and the sum of the south measurements may end up being a faction of an inch or so *less* than the sum of all north measurements. The same thing could happen with a comparison of the sums of the east and west measurements. And the sum of the angle measurements might be off a few seconds or minutes. Why does this happen?

There are many reasons, but first let's consider why the sides could be off:

- As we see in Figure 20, even when a tape is held taut, it will always have a slight sag due to its own weight. Note the following illustrations. This means that when the sag is measured, it will be a bit longer than the horizontal measurements.
- If a 100-foot tape is being used to measure a plot a thousand feet long, for example, it is not always possible to butt the beginning of the tape precisely at the point where it ended before when taking successive measurements.
- Hills and valleys may make it difficult for a horizontal measurement to be made.
- Over very long boundaries, even changes in temperature may affect the length of the tape enough to record a slight variation instead of the true measurement.

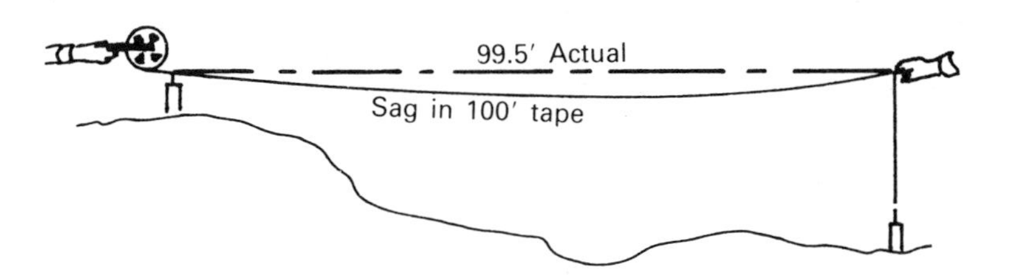

Figure 20 *Illustration showing how 100-foot tape can sag and thus line of 99.5 feet could be read as 100 feet.*

The angle totals may be off, too, because of its being impossible to sight the transit minutely enough to arrive at a precise reading. Or perhaps it is impossible to center the transit over the exact center of the corner monument, thereby making the measurement off a fraction of a degree or minute.

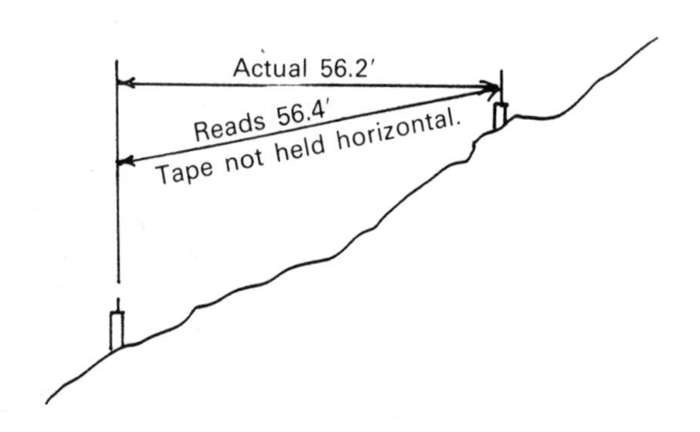

Figure 21 *Only horizontal measurements are true measurements of distance.*

Accuracy vs. Precision

When we speak about accuracy in surveying, we mean that the final drawing of a parcel will show the outlines of the parcel of land correctly. If a stone wall is shown on the plan as the east bound, or if an iron pipe is found at the end of a barbed wire line, it can be assumed that these are the bounds of that parcel. Lines and markers are shown to be correct in all respects in accordance with all available data that the surveyor was able to find after an honest and painstaking research job. And unless some unknown and unfiled deed suddenly comes to light, a court will accept the boundary lines as the true lines even though there may be slight deviations in boundary and angle measurements so the survey does not technically represent a "perfect" closure.

As compared to *accuracy, precision* refers to the degree of exactness of the measurements of boundaries and angles of the parcel. Another way of stating this is, how great is the "error of closure"? As we have said, it is well nigh impossible under field conditions to make a perfect closure; there will most likely be some error that has crept into the measurements regardless of how careful the surveyor is and how good the equipment is.

Time vs. Cost

Let's assume that a surveyor can come up with a closure plan that is 95% precise. Is this good enough? Or should he strive for one that is 99% precise? (In the language of surveyors, this would be noted as an error of closure of 1′ in 5000′, 1′ in 10,000′ or 1′ in 15,000′; i.e., when they close the survey mathematically, they would not be off more than 1′ in 5000′, etc. . . .) Today's surveyors have, in most areas, changed over from the transit and tape to the theodolite and electronic distance measurer (see this equipment in Figure 40). As equipment allows for more precise surveys, and as computers provide swifter means of checking work, states are setting more exacting standards to which the surveyor must adhere. Nevertheless, the surveyor could spend many, many hours on the survey to minimize as much as possible the error of closure. All this would increase the cost, perhaps manifold, depending on the percentage of precision the client demands. But is it worth the added expense? How does one decide?

The decision is an economic one. A client with a 100-acre farm or woodlot in a rural area is not going to demand 99% precision. Why do so, especially when the abutters have approved the boundaries? Why go to all the added expense if the purpose of survey is to confirm certain stone walls as the property lines? The surveyor can then make up a drawing showing these as the actual bounds, and the deed can then be improved to read something more than "by land of Jones; thence by land of Smith." It will confirm these lines with the neighbors and prevent problems should a neighbor suddenly die or sell to others.

But what if the parcel is in a city where an acre of land is worth a million dollars? In this instance every square inch of land is important, and the client will be willing to pay for a higher percentage of precision of the closure than will the client who owns the farm or woodlot. If the client is selling the city parcel, he will want to get the maximum price for it; so every square inch is important to him. If the client is buying the city property, he will not want to pay for a square inch that is not his; so every square inch is important. And for those parcels that fall in between the 100-acre farm or woodlot and the city parcel, each client requesting the survey must determine whether the cost of a more precise closure is worth the price.

Area Standards

In summary, then, the degree of mathematical closure (or of minimal error of closure) that may be required for a particular survey can only be determined by you and the local codes. Should you desire nothing more than a tape-and-compass survey to check the amount of acreage for tax purposes, that is one thing; but should you desire a precise job for purposes of a subdivision, the cost will be considerably more. A tape-and-compass survey will be inadequate for purposes of subdivision, for an error of closure of that extent could not be tolerated in the layout of lots and streets. Here is where the more sophisticated electronic equipment is needed for greater accuracy.

Chapter 6 *Deeds and Deed Descriptions*

Description

What is a deed? What is it supposed to do?

A deed is a legal document setting forth the fact that a parcel of land has been conveyed to another party. A deed should contain the following elements:

- Date of the land transaction.
- Identification of the person who conveys the property to another, who is also identified.
- Description of the property conveyed in specific terms that will give the precise bounds, corners, and number of square feet.
- Location of the parcel on the face of the earth.

Its precision should also assure that the bounds and corners of the property are in agreement with the deeds of the landowners who abut the parcel in question. If this is not done, then there could always be a question of boundaries which could arise when those abutters survey their land.

Our main concern in this chapter is the legal description of the property being conveyed.

History of Deeds

A landowner was heard to remark, "My deed gives a good description of my property, together with all bearings and dimensions you'll need to survey it."

Unfortunately the above statement is not, repeat not, true in far too many cases. There are numerous reasons for this. One pertains to the background of recorded deeds.

Unless we live in a recent development, the chances are that the parcel of land that we own, or wish to own, has not changed hands very many times in the past two hundred years. Originally the parcel may have been surveyed and plans and/or deeds written up accordingly. As time progressed and this land was split up among families or a parcel was sold off, the deed for the sale was written in reference to the original deed. This was especially true in New England after the surveyors headed west about 1835 or so to help lay out the new states and railroads. It was simpler to refer to the original deeds when land was sold; hence we find reference to "by the land of Smith" (who probably died in 1900 or before), and "by the land of Jones' heirs," etc. Deeds written even today very frequently carry this old method in spite of the fact that the original Smith and Jones have long since passed away. Surveyors must therefore trace the land sales back through the records in an attempt to find some deed in the line which will give them information as to the originally-intended bearings and dimensions.

Meaningless Descriptions

Visualize, if you will, a surveyor being handed a deed referring to the original owners and trying to find its location on the surface of the earth. Even though it may have been handed down from father to son a couple of times, and even though each generation knew exactly where it was located, try to imagine a wife who suddenly finds herself to be a widow. Her son has left home for greener pastures, she has never walked the lines, and the deed description means nothing to her. There is no one to walk the lines with the surveyor, and neither he nor the widow know if perhaps part of the land mentioned in the deed has been sold off, or if there may be other deeds she has not yet found that relate to the property. But now the surveyor must research it and finally fix it on the face of the earth.

Problems have occurred in this respect when neighboring parcels of land are purchased. This abutter then claims part of this widow's land, and due to the flimsiness of the deed, the argument often stands up in court. Courts invariably have no other recourse than to split the land between the two abutters with each taking half of the land in dispute.

This is one of the most impressive arguments for keeping deeds updated, for as stated before, although you may be sure where the boundary line is, and sure your longtime neighbor knows, your neighbor may sell and leave town and your new neighbor may question the boundaries at some later date.

Subdivisions of Land

A similar problem very often arises regarding lands that have been subdivided. Even though the original plans for the subdivision have been filed with the registry, complete with bearings and dimensions and the subdivision's lot number of the property in question, corner markers are often missing. This is partly due to the fact that until more recent times, corner markers were not required to be permanent. But even when they have been set in the ground, they have often been nothing more than pipes that were removed when the lawn was graded, or knocked down and never replaced. Or they were replaced but not replaced accurately. A hedge between a house on the property and a neighbor's house may have become accepted by both the owner and the abutter as the property line. But is it really?

Angled Lines

When towns were originally laid out in a grid system by the selectmen, what were called "range" lines ran north and south, and "base" lines ran east and west. The result was a division of the township into "checkers." See Figure 22.

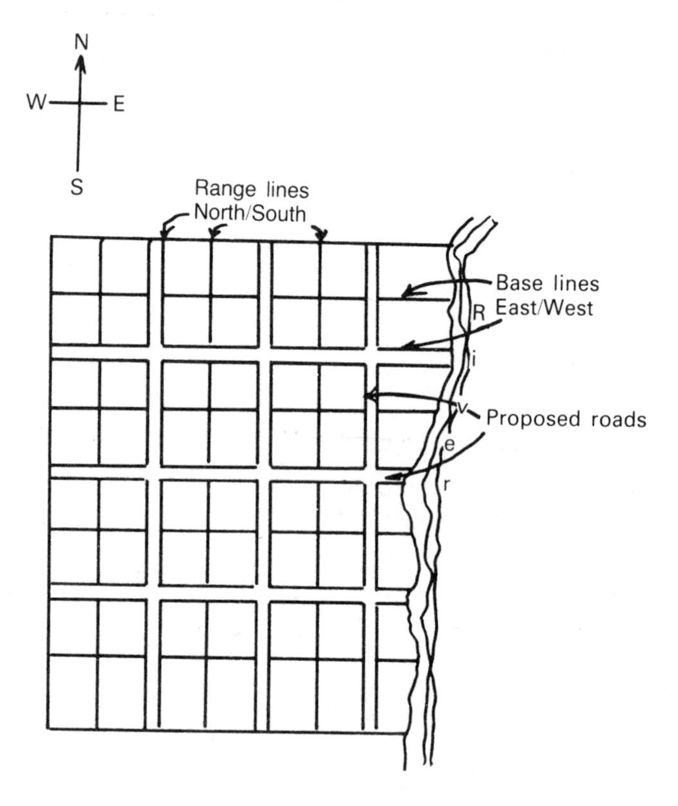

Figure 22 *Range and base lines for layout of town.*

It was proposed that the range and base lines would be roads, and the checkers would therefore be rectangles, since the range and base lines ran perpendicular to each other. But due to the courses of rivers or due to swamps or high hills, the roads could not always follow the original range and base lines. It became necessary to go around the problem areas. See Figure 23.

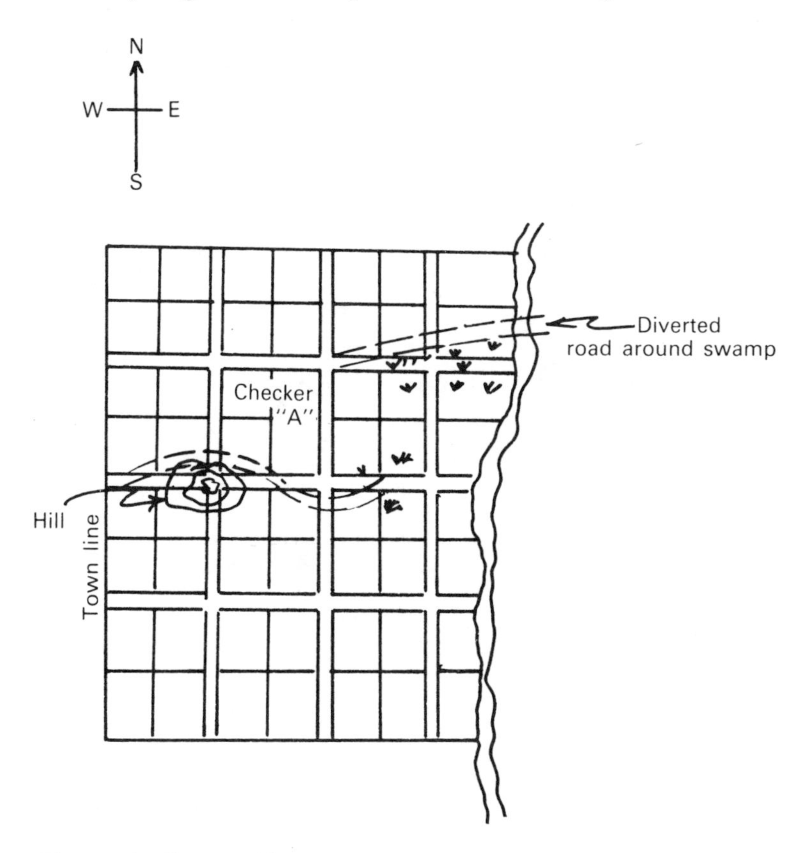

Figure 23 *Swamps, hills, and other natural obstructions often prevented range and base lines from being used as roads, thus requiring roads to go around such obstructions.*

When this happened, roads became angled, and new problems were born. Figure 23 shows how the actual road became angled in relation to checker "A."

Note the parcel in the lower left-hand corner of checker "A." Suppose now that a buyer wants an acre of land that is as long as it is wide. A square with all 90-degree angles would be 208.71 feet long on each side for the 43,560 square feet constituting an acre.

By not surveying this land and by not taking into account the angle of the road involved, the owner has left himself open to problems. One is that he will require added depth along the left side due to the angle of the road, or he would not get a full acre of land. For instance, if the angle should happen to be 120 degrees and not 90 degrees in the left front corner of the parcel, it would be necessary to go back along the left line for a depth of 240 feet to make up the square footage required for an acre of land. (At 209 feet along the side, he'd have only 37,830 square feet.)

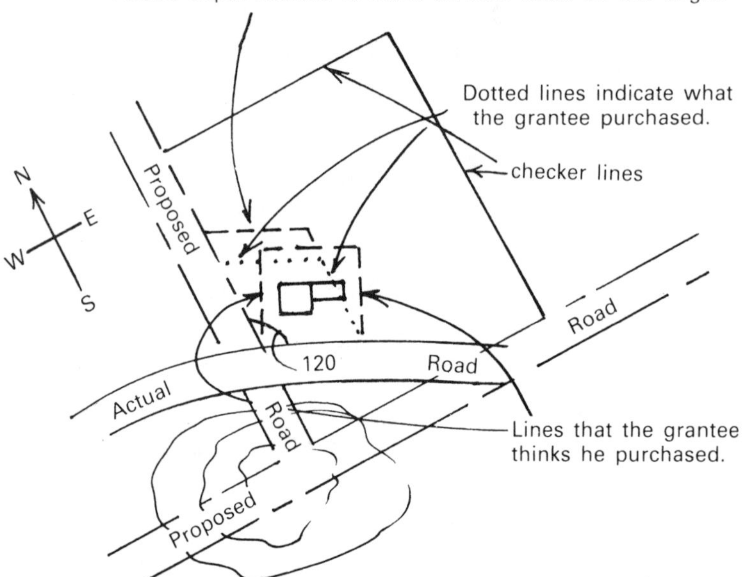

Figure 24 *Problems arose when road curved to avoid obstruction and owner did not have his land surveyed before purchase.*

The other problem is the owner has assumed that 209 feet along the road is ample frontage. He sites the building in the middle of what he believes is the parcel an ample distance off the road and commences to build. Suddenly it is discovered that the building hangs over into the grantor's land. Why? Because of the wording of the deed. Figure 24 shows the line the buyer "thought" existed—one whose easterly corners were 90-degree corners. In other words, the corners were not spelled out, and it was discovered too late that what was "thought" to be a perfect rectangle was in fact a parallelogram.

This same problem arises in instances where a house is on a parcel of land with a road that is later shifted. The then-owner sells the land out from around the house with the deed reading so many feet *along the road.* This footage is arrived at by pacing down along the road until the building is passed. The land is then surveyed, and it is found that the property lines runs through the building because no one took into account the angle of the fencing or line on the right-hand side. Figure 24 again illustrates this problem.

Solution

Such deeds written and filed throughout the years leave no alternative other than to attempt to settle the differences between a landowner and a neighbor as they arise. The simplest solution is a quit-claim deed from the neighbor giving the landowner the agreed-upon portion; it will clear the records insofar as settling the lines, but a survey of the new agreed-upon lines is the final answer in order that the new deed can be written around the proper and correct bearings and dimensions.

We mentioned deeds which a widow might find herself possessing. These can only be searched down through the years piece by piece as you would put together an intricate puzzle. This means many hours of research—research that may require wide areas of examination before the final boundaries of the parcel can be defined. Even then, this may require agreements being reached with numerous abutters, and with their abutters, to be assured that all problems are settled once and for all. Such cases can be settled only by the local land surveyor working closely with the attorneys who are involved with the deeds. It is a sad fact that occasionally the expense of all this research adds up to a great deal more than the final parcel is worth.

Writing Deed Descriptions

That National Geodetic Survey, which is run by the U.S. Department of Commerce, has over the past century and a half imbedded over a million bronze markers in concrete or bedrock in different parts of the United States and its possessions to indicate official latitudes, longitudes, and elevations. These markers are for the use of engineers, surveyors and mapping agencies as ties to their work. It is hoped that eventually there will be sufficient numbers of these markers so it will be possible to use them as reference points in locating properties on the face of the earth.

For instance, think of the road map you use when you want to locate a town. You look in the index of towns, find the town desired, and beside it find a letter and number; i.e., opposite Jamestown, Rhode Island, might be "B-4." You find the letter "B" in the right hand vertical column of the map and the number 4 in the bottom horizontal column. Where these two intersect is Jamestown. Now visualize a similar system of reference numbers on the face of the earth using National Geodetic Survey monuments. You can see where a so-called coordinate system could be set up to tie your land down. Unfortunately, there are only a few of these available at this time.

Earlier we identified the elements that a deed should contain in order to locate a parcel at a precise spot on the face of the earth and to describe fully the bounds and the area the parcel contains. Now let's use the example of the 100-foot-by-100-foot square lot used in Chapter 5 on Mathematical Closure and describe this as it should be written into a deed.

We'll also give it some ties to a house which is set upon it. These can be used to locate the points on the face of the earth. This information, along with the state, town, and street names, will give the surveyor the approximate location of the parcel—something more than just the names of the abutters—i.e. "by land of Jones," etc.

Let's dress up our 100-foot-by-100-foot parcel a bit before we write a description of it. See Figure 25.

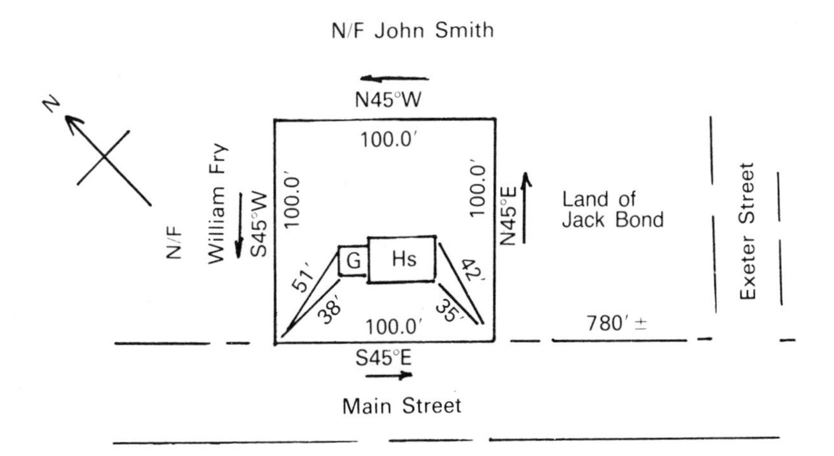

Figure 25 *This parcel provides basic information for the deed that follows.*

Total Deed

The following will indicate not only the legal work required to make this a binding deed but a description of the land which will place it on a precise spot on the face of the earth.

> Know all men by these presents, that I, John H. Martin, of Jamestown, Bristol County, Rhode Island, in consideration of one dollar and other good and valuable considerations paid by Jachon Hill, of Dumont, Thames County, South Carolina, the receipt

whereof I do hereby acknowledge, do hereby give, grant, bargain, sell, and convey unto the said Jachon Hill, his heirs and assigns forever, a certain parcel of land with all buildings thereon, situated on the northerly side of Main Street in said Town of Jacksonville, Rhode Island, County of Bristol, bounded and described as follows, to wit, Beginning at a point in the northerly line of Main Street in the town of Jacksonville, Rhode Island, at a stone bound set in the ground, said stone bound being approximately 780 feet from the intersection of Main and Exeter streets, at the southwest corner of land now or formerly of Jack Bond; said corner also tied to the house foundation by a tie of 35′ to the front corner of said foundation and 42′ to the rear corner; existing foundation being 24′ wide; thence running along land of Jack Bond N 45° E for a distance of 100 feet to a stone bound set in the corner at land of John Smith; thence turning and running along land of John Smith N 45° W for a distance of 100 feet to a stone bound set at the corner of land now or formerly of William Fry; thence turning and running S 45° W for a distance of 100 feet along land of Fry to a stone bound set in the corner at the intersection of the line of the northerly side of Main Street and tied to the present 22-foot-wide garage foundation by 38 feet to the front corner and 51 feet to the rear corner; thence turning and running along said Main Street S 45° E for a distance of 100 feet to the point of beginning. Said parcel of land containing 10,000 square feet, and all bearings refer to magnet north 1978.

Being the same premises conveyed to Jon Field by Ralph Johnson on September 10, 1936, and recorded in registry of deeds of Bristol County. (At this point would also be stated the book and page number.)

Is This Sufficient?

Reviewing the deed, it would appear that all the required dimensions and bearings with which this parcel of land could be laid out are included. We also have the tie from Exeter Street which places it on the line of Main Street. We then have the ties from those two front corners which will set the two corners exactly on that line of Main Street. Had a mylar or ink drawing been done to file at the registry of deeds, we could have referred to this, and the ties could have been shown on it. If a survey plan had been filed, then this would have been noted in the deed as "in accordance with Plan Book _____, page _____."

The question naturally arises as to what would happen if the street on which the starting point is located were moved or if the house or garage structures were changed, thereby changing the ties? How could the points be located then? The answer is that it is the owner's responsibility to be aware of these changes and establish new tie points that can be recorded. We should never forget that ownership carries the responsibility of maintenance, and preventive maintenance is part of maintenance.

Various Types of Deeds

The deed used as an example is a standard type of deed to convey land from the grantor to the grantee. But there are other kinds of deeds that are used in connection with land transfers and land problems. The two most familiar are the *warranty* deed and the *quit-claim* deed. There are others such as *bargain* deed, *sales* deed, *contract* for deed, *sheriff's* deed, and *tax* deed, but these are all special types for specific purposes, and an attorney should be consulted for uses of these.

Quit-Claim Deed. This is a deed of conveyance, intended to pass any interest, claim, or title which is held by the grantor on the specific premises but not professing to contain any warranty of title and interest by others; i.e., quit-claim deeds convey all title and interest that a grantor may hold on his property, but they cannot guarantee that someone else may not step forward with a claim of some sort on the land. As we have noted, a quit-claim deed is commonly used to confirm boundary agreements between abutters.

KNOW ALL THESE MEN BY THESE PRESENTS, THAT

> I, John Martin, of Jamestown, Bristol County, Rhode Island, in consideration of one dollar and other valuable considerations, do hereby give, grant, bargain, and quit-claim all my rights, title and interests in land located in the town of Jacksonville, Rhode Island, on the northerly side of Main Street to the said Jachon Hill, his heirs and assigns forever, Said parcel being located on the northerly side of Main Street and containing approximately 10,000 square feet and bounded and described as follows: A certain tract or parcel of land situated in Jacksonville, Rhode Island, bounded northerly by land of John Smith; easterly by land of Jack Bond; southerly by said Main Street; and westerly by land of William Fry. Said parcel being the same described in deed to Jon Field by Ralph Johnson on September 10, 1936, and recorded in Bristol County registry of deeds in Book ______, page ______.

Warranty Deed. In a warranty deed the grantor warrants clear title with warranties of seisin (legal possession), right to convey, and freedom from all encumbrances and with defense of title accordingly.

From this definition you can see that a warranty deed is the preferable one, since it leaves the grantee free from responsibility insofar as problems and questions regarding the land being purchased. With a quit-claim deed, the grantee takes the chance that other ownership could arise.

But regardless of the type of deed, it is important to determine that it spells out the complete parcel and places it precisely on the face of the earth.

Visualize again the road map with letters up and down the sides and numbers along the bottom. We'll call the vertical lines the "y" axis and the horizontal ones the "x" axis, and assign them number values in describing a parcel of land with coordinates. For simplicity call the Geodetic Survey point 1000 for its "y" axis and 1000 for its "x" axis. By doing this we now have good "ties" or starting points to pin this parcel of land down to the face of the earth.

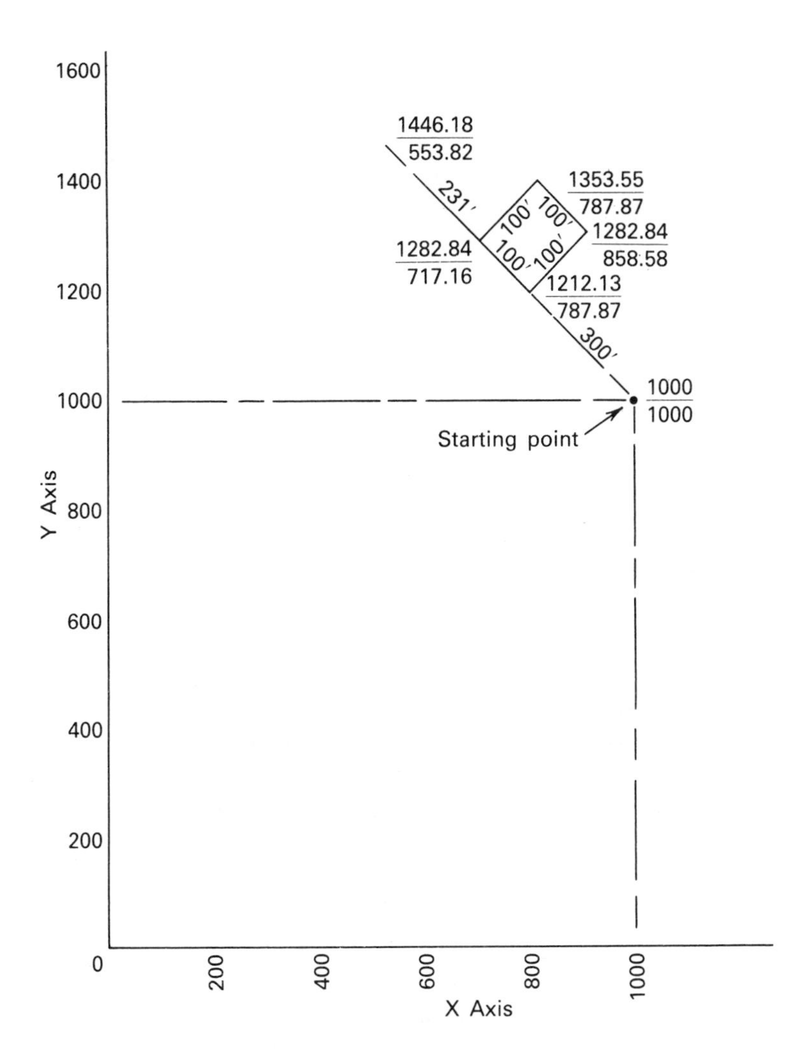

Figure 26 *Illustration showing coordinate method of laying out parcel of land.*

Now let's see what it would look like in the deed. It is possible to write a deed with just those coordinates for the corners given. They would tie this down, and a surveyor could compute from those x and y figures just which route he was to take, i.e., which bearing and dimension would take him to the various corners of the land. The deed might read: "Starting at a point which has coordinates x-1000 and y-1000, which is a Geodetic Survey point on Main Street, being a brass cap set in concrete; thence to a point having coordinates of x-787.87 and y-1212.13, which is a stone bound and the point of beginning of the parcel to be herein described; thence to coordinates y-1282.84 and x-858.58, which is a stone bound at land of Smith . . . etc." This would continue around to the various corners and even tie into the coordinate point 231 feet up along the street.

Now let's look at a normal description which would give the bearings and dimensions and use those coordinates for ties to the earth only:

STATUTORY SHORT FORM OF WARRANTY DEED

I, John Smith, of Evertsville, Bristol County, Rhode Island, for consideration paid, grant to Leslie Schwartz of Evertsville, Rhode Island, with warranty covenants the land in Evertsville, Rhode Island. A certain lot or parcel of land with buildings thereon located on the northerly side of Main Street in the town of Evertsville, Rhode Island, and bounded and described as follows: Beginning at a stone bound on the northerly side of Main Street, which bears N 45° W at a distance of 300 feet from a Geodetic Survey marker of brass and set in concrete with coordinates of y-1000 and x-1000; thence running along land of Jack Bond N 45° E for a distance of 100.00 feet to a stone bound at land of John Smith the Grantor, and said bound having coordinates of 1282.84-y and x-858.58; thence running along land of Grantor Smith N 45° W for a distance of 100 feet to land now or formerly of William Fry to a stone bound set in the ground; thence turning and running S 45° W along land of Fry for a distance of 100.00 feet to a stone bound on the northerly side of Main Street; said point being S 45° E 231 feet from Geodetic Survey marker with coordinates of y-1446.18 and x of 553.82; thence turning and running along said Main Street for a distance of 100.00 feet to the stone bound and point of beginning. Said parcel containing 10,000 square feet, and all bearings refer to True North. To have and to hold, the aforegranted and bargained premises with all the privileges and appurtenances thereof to the said Leslie Schwartz, his heirs and assigns forever. And I do covenant with the heirs and assigns, that I am lawfully seized in fee of the premises, that they are free of all encumbrances; that I have good right to sell and convey the same to the Grantees to hold as aforesaid; and that I and my heirs, shall and will warrant and defend the same to the Grantees, the heirs and assigns of the survivor of them forever, against the lawful claims and demands of all persons.

IN WITNESS THEREOF, I the said John Smith, said Grantor, relinquish and convey all right by descent and all other rights in the above described premises, and have hereto set my hand and seal this twenty-sixth day of February in the year of our Lord one thousand nine hundred sixty.

Signed, sealed, and delivered
in presence of

______________________	______________________
Clyde E. Anthony	John Smith

State of Rhode Island, Bristol ss. February 26, 1960 the above named John Smith appeared before me and acknowledged the foregoing statement to be his free act and deed.

Before me, ______________________ Clyde E. Anthony   Notary Public, Commission Expires April 30, 1980.

The Wording of Deeds

There is no single wording that a deed must have, either for its beginning, description of property, or its ending. Convention seems to have a role in some parts of the country; lawyers will also use different ways to state the elements a deed should contain.

Main Intent

The main intent of a deed is to put forth a binding contract between two parties and assure the grantee that what is being purchased is free and clear of all encumbrances. To some degree, the deed is subject to the legal profession's use of words. Such use can sometimes confuse the layman, and so we should take careful note of these contracts and assure ourselves that they not only bind the land to us, but that they also describe the parcel in language that we can well understand and could locate precisely on the face of the earth. Because of all the entanglements that can come about due to conflicts of abutters' deeds, our deed, at least, should contain no uncertainties.

Often we are advised to purchase land and to worry about the lines at some later date. Land is a dear item on today's market. We do not buy a car without first becoming knowledgeable about its horsepower, gas mileage, comfort, and other specifics. We should be even more knowledgeable about the land we wish to purchase.

Chapter 7 Marking Your Land

There are two very important things to keep in mind to protect your parcel of land from intrusion:

- Corner markers must be clearly recognizable at each and every corner of your land.
- These corner markers must be maintained and protected from being lost or encroached upon.

Corner Markers

Figure 27 shows the many kinds of corner markers, and there are many other kinds that are used from time to time: pipes, wooden stakes, trees, and even piles of stone. Only the imagination seems to limit what may be used; e.g., in North Berwick, Maine, there was once a deed that identified a corner marker as a particular "hole in the ice," for at that point the small body of water had a spring underneath, and ice never froze in the winter. But corner markers should be permanent and easily identified.

Some towns have adopted codes restricting the types of corner markers recognized as legal, but where that is not the case, we recommend using a reinforcing rod, which is available at most steel supply houses. After it is driven into position, the earth surrounding it should be dug away to a depth of twelve to fifteen inches; concrete should then be poured into the hole around the rod to make it permanent.

If there are granite quarries in your area, they can supply excellent markers. So can manufacturers of concrete products; they can fabricate a four-foot tapered concrete corner marker with a piece of reinforcing rod down its center and protruding at the top for centering on the corner boundary.

Perambulation of Lines

When our forefathers laid out the towns and placed markers at the town bounds, they foresaw the problems that could occur should these markers not be maintained. In the records of the state of Maine we find the following law, which contains good advice:

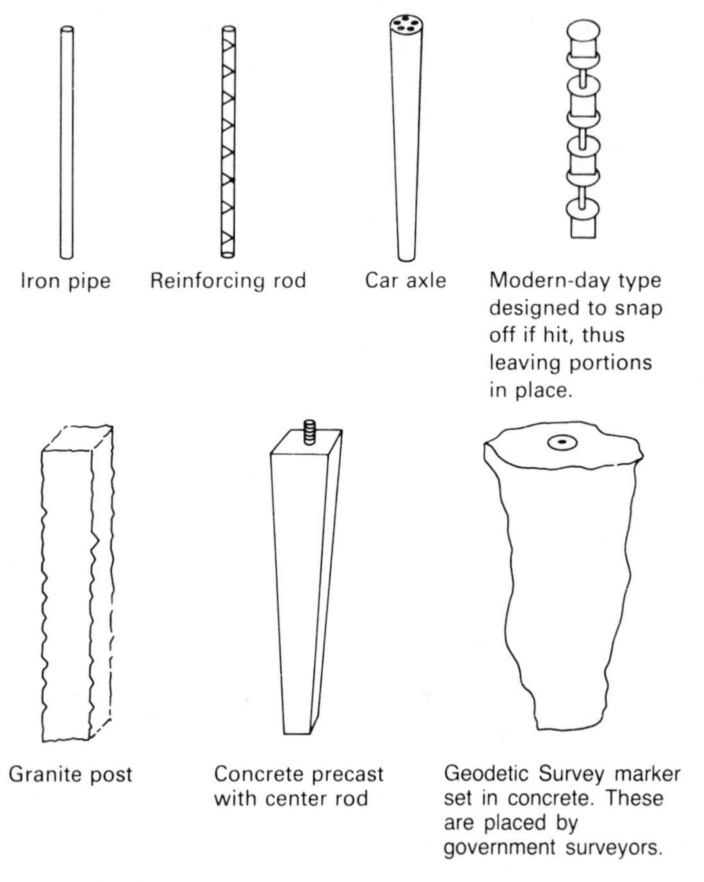

Figure 27 *Various types of commonly used markers.*

> "Boundary lines between municipalities shall be run [sic—"walk" is intended] once every five years in accordance with the following provisions:
>
> "Municipal officers shall give 10-day written notice to abutting towns that they intend to perambulate [sic—"walk" is intended] the lines . . .
>
> "that the expense shall be borne by each of the abutting towns . . .
>
> "that recordings shall be made in the town books that this has been conducted . . ."

The law goes on to describe types of bounds and their durability:

> "Municipalities which have perambulated their lines and have erected stone monuments, which protrude at least 2 feet above the ground, at all angles, edges of highways, and bodies of water, which the boundary line crosses, or which serve as boundary lines, are exempt from perambulating the lines, except once every ten years from the time that the monuments were first erected. *This perambulation is for the purpose of replacing those which have been misplaced or destroyed.* (Author's italics.)

The story goes that our forefathers had just such a tradition as this for family landholdings. The families on either side of a property line would gather and would walk the line. As they reached a point on the line that a father wanted his son to remember, he would "bump" the son against that point to "impress" it on him. It is also said that the boy would receive a small cake or cookie as a reward. This is a good custom and one that would be well to carry on!

Individual Parcels

What does this all have to with our own little parcel of land? Nothing, if we are one of the fortunate few who have bounds that we are sure cannot be moved in any way and that can be seen by us each day as we roam the yard, mow the lawn, or walk to and from the car.

Otherwise we, too, should be walking our lines now and then to keep them well marked—pipes well painted, granite posts clipped around so grass and weeds will not obscure them from view, and any line of trees that we are sure is a boundary line slashed and marked with a touch of fluorescent spray paint.

When selecting trees to be marked, choose those that are at least four to eight inches in diameter and ones that are not likely to be selected for lumber and other use. Naturally, some of the trees will be directly on the line while others will be a short distance inside the property line. Those on the line should be marked on both sides where the property line goes through the tree. For those just slightly inside the line, mark them twice—one on each side of an imaginery line running through the tree perpendicular to the property line. Never select trees farther from the line than the length of an arm plus an axe handle, and in those instances the mark should be directly opposite the property line. All marks, whether blaze marks, hack marks, or other, should be approximately five feet above the ground and preferably five to six inches in diameter.

Care should be taken, however, that you do not destroy any of the old marks, since you would be invalidating their proof.

And most important, neighbors should be constantly informed as to what is being done to preserve and maintain markers.

Areas of woodland with nothing more than stone walls or barbed wire fences to identify bounds are very susceptible to intrusion; so such parcels should be perambulated with considerable frequency to assure that monuments are kept in shape.

No Monuments

In those cases where there are no monuments, our best advice is to set up a meeting with abutters as soon as possible and settle on a corner (if there is nothing in the deeds or records to prove one) and set in a permanent marker. A sketch of the area where the markers are placed, together with any tie dimensions, can then be drawn and copies made for each abutter. Signatures of each will assure that this was agreed to by all. Such signatures should be those that appear on the deed.

And whenever old markers are replaced with new ones, the abutters should also be notified and approve the corners. It is good to remember that any such corner is not only the corner of your property but also the corner of your neighbor's property.

Power companies often place their poles on the points where your corner markers have been placed. Placing a pole on the corner assures the company that they are not stringing wires over anyone else's property. See Figure 27A. This is another reason to be sure that your corners are clearly marked.

Robert Frost may have been ridiculing small-mindedness when he said, "Good fences make good neighbors." There is no doubt, however, that "boundary markers make good neighbors."

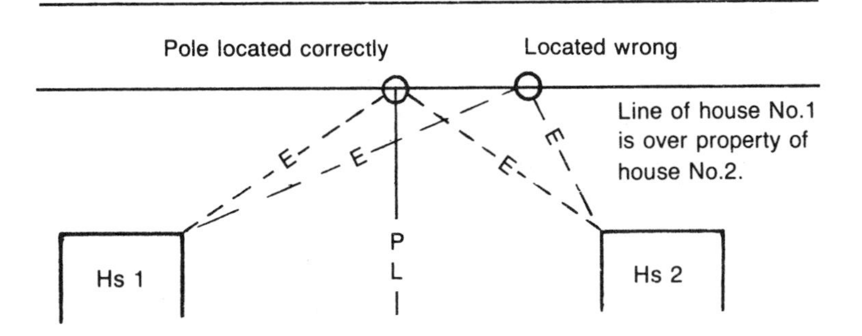

Figure 27A *Correct placement of utility pole.*

Chapter 8 *Changing Corner Markers*

Whenever corner markers become worn and deteriorated, they should be replaced with new and more durable ones. There are several ways to do this and still place the new marker in the precise location of the old one.

The Right of Abutters

As we have noted before, a boundary marker is not only the property marker of the owner of the parcel of land in question, it is also the boundary marker of the neighbors. Thus it is a common bound between the parcel owner and the owner of the land abutting the parcel.

We have already said that when markers are set, it is important to get the approval of any persons who abut your property in order to eliminate any later questions of property lines.

The same is true when one wishes to replace already positioned corner markers with new ones. Once an abutter becomes accustomed to a certain type of monument for a corner marker, he could well question the motives if that marker were suddenly replaced with another type without his knowledge. Since corner markers represent common-interest points, it is clear that they should have common approval of design and location.

Methods of Replacing

Whether we are changing the land surveyor's stake at the corner, restoring the rusty iron pipe, or just replacing that old car axle, our major concern is that we reposition the new marker in the same location as it was when all abutters agreed on it. This can be done by various methods.

If we are out in the woods where we have trees in the vicinity, we can drive a nail into each of two trees and take dimensions of the distances from the corner marker to these nails. See Figure 28.

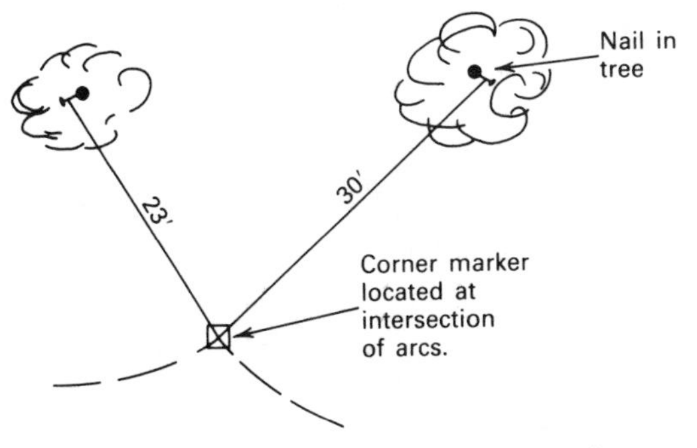

Figure 28 *Locating corner marker by use of trees.*

Once we have dug the hole and approximately located the new marker, we can then accurately position it from these tie dimensions.

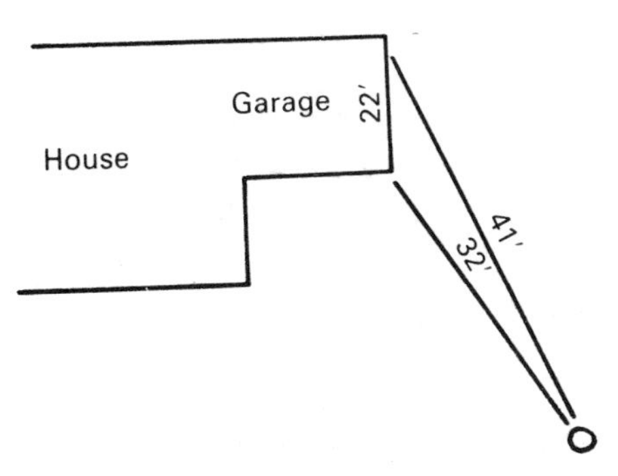

Figure 29 *A corner marker tied to the foundation of a house.*

If, instead, we have a survey plan showing ties to a house, we can use these to locate the new corner marker. See Figure 29. (Note that in most instances the ties will have been taken from the house foundation vs. the shingles or siding. This prevents errors should the siding ever be changed.)

Should you find yourself in the middle of a field with no permanent fixtures to tie to, you can use the stake-and-string method as follows: See Figure 30.

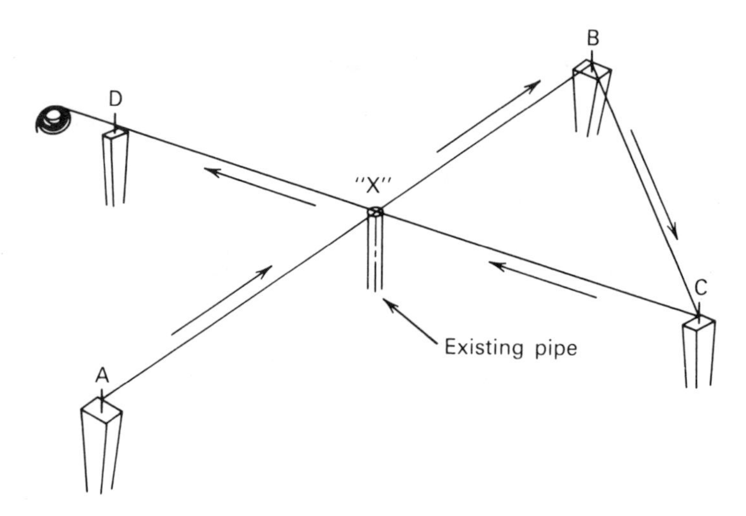

Figure 30 *Use of stakes in ground to provide method of locating precise point where new monument will be set when old one is removed.*

- At the corner in question, drive a stake into the ground three or four feet from that corner marker but leave it exposed high enough to clear the top of the marker you are replacing.
- Next drive a nail into the top of that stake and tie your string around it, pulling the string taut over the centerline point of the top of the marker (X in above illustration) to three or four feet past it, i.e., a to b.
- Now while you hold the string in a straight line over the corner marker, set a stake and drive it under the string (b). With the

stake set, pull the string taut over the center top of the corner marker (X) and over the second stake (b). Drive a nail into the top of the stake and wrap the string around it and tie it.

- Do the same thing again, but with the second string at right angles to the first one, i.e., c to d. Be sure the second string passes over the precise center of the corner marker (X).
- Now release the string from the nails, and, leaving the stakes in place, carefully dig the hole required for the new monument.
- With the hole dug, set the monument in it and, after retying the strings, adjust it so its precise center will be under the point where the two strings cross. Now carefully backfill and make final adjustments prior to tamping around the new monument.

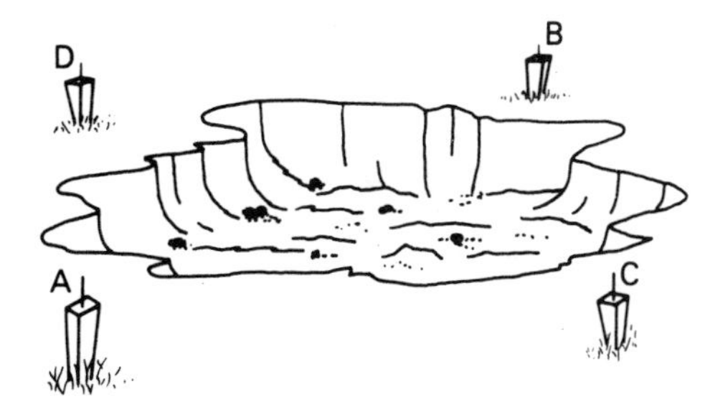

Figure 31 *Old monument removed, but stakes must remain in undisturbed positions until new monument is set.*

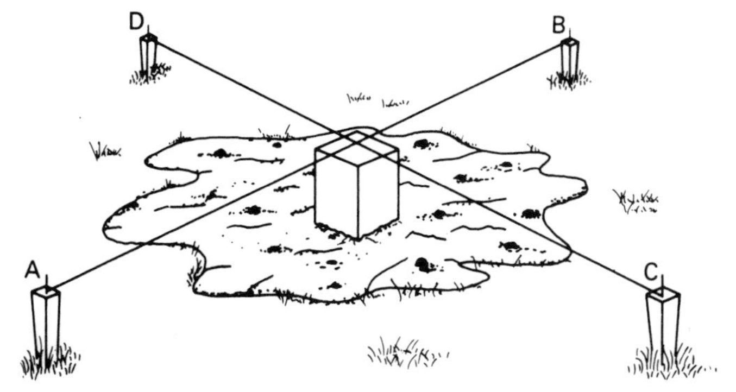

Figure 32 *The exact center of the new monument must be directly under the point where the lines from the four stakes intersect.*

Centerlines and Boundary Lines

It is commonly understood that monuments are placed on the centerline. Moreover, in the case of a stone wall, we think of its centerline as being the boundary line, and accordingly, a corner would be under the centerline. Nevertheless, sometimes this is not the case; so it is important to read the deed carefully.

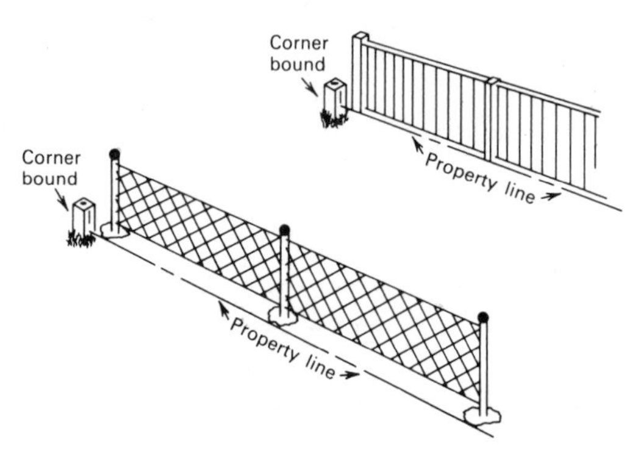

Figure 33 *All fencing material should be on property of the owner of the fence.*

Modern Fences

When today's modern fences, such as a wooden ranch-style fence, are installed, they are usually placed right on line. To say "right on line" means they will not infringe on the abutter's property but will come to the very edge of the owner's property.

But the concrete footings of a chain-link fence will be situated so they are completely on your land. This will necessitate having the fence a few inches inside the property *line,* but it assures you of no problem with your neighbor tampering with the footings of the fence.

When buying or selling land, these points should be studied carefully to determine who actually does own the fences.

Chapter 9 *Lawyers and Legal Considerations, Including Title Search*

Role of a Lawyer

It is the role of the lawyers, attorneys, or counselors to defend or prosecute causes in the courts of the United States. Their business is also to provide assistance or legal guidance in matters pertaining to the law.

In property transactions, their assistance is desired for title searches and preparation of deeds. Although there is no law requiring that a lawyer write up deeds, it is recommended that lawyers prepare them.

Buyer and Seller Lawyers. A common question during a land transaction is whether it is necessary to have two lawyers handling the transaction—one representing the buyer, the other representing the seller. Good clear title to property can be provided only if the buyer has his own lawyer—not the same one who is working for the seller.

Deed Writing. An attorney familiar with land transactions and modern-day problems that arise will be able to advise if the deed of the previous owner is sound or whether it should be improved upon by a field survey. The fact is, however, that few attorneys understand fully all of the modern-day problems involved in the description of a parcel.

Legal descriptions are necessary, and our deeds must be written in language that will stand up in a court of law, but today's world also requires that a deed provide full data essential to locate a parcel on the face of the earth.

This is where the cooperation of the land surveyor and the lawyer can benefit the buyer and prevent future problems. *There is no possible way that a precise description of a parcel of land can be written without knowing the plan of the land surveyor.* Furthermore, because of major changes in equipment and technology, along with added understanding of problems that must be corrected in older surveys, old survey plans that are on file now often are not capable of providing the necessary and precise information. For example, if the last survey of your land was done with compass and chain, it should be resurveyed to meet today's standards. (See Chapter 12 for a discussion of past, present, and future methods of surveying.)

Even parcels of land which have been recently surveyed should be brought up to date from time to time to reflect new requirements for greater precision as land values increase.

Boundary Disputes When property-line disputes occur, a land surveyor should be contacted at once. In the majority of cases, he will be able to research all data available regarding both parties involved, discuss the problem with the abutters and, in most cases, reach an agreement with them and then do a survey if necessary. The surveyor can advise if it is necessary to obtain a lawyer for possible changes in the deeds.

If the lawyer is called in first, his immediate approach is usually to argue, as he should on his client's behalf. The lawyer than calls in a surveyor to lay out the land on the face of the earth as he wants it in the deed, which immediately riles the neighbor, who then hires his own attorney and land surveyor.

There are now two lawyers striving to prove their clients' deeds are right and two land surveyors working from two directions to lay out two separate parcels of land with vague bearings. Should the land overlap, the end result will be a court calling for the parties to "split the difference."

Legal Sequence in Surveying

Over the years, the courts have set priorities to the sequence in which legal considerations involving surveying should be undertaken. These considerations and their priorities are as follows:

Fixed monuments, such as a watercourse or a ledge mark.
Man-made monuments, such as a stake, iron pipe, or concrete marker.
Distance, such as the length of a course or a line.
Bearing, the heading or bearing of a line.
Area, the square footage or the acreage.

Monuments. Monuments of the fixed type come first, since there is little chance that they can be tampered with.

Man-made monuments come next, and care must be taken that they are as originally placed or that there is sufficient evidence to back up any changes in their location. To support such changes, the landowner must take care to keep his records up-to-date. Was a neighbor in attendance when a monument was substituted for that old rusty pipe?

Distance. The lengths of the boundary lines are considered next, and in this respect we must take into account the degree of precision of today's measuring equipment vs. those of yesteryear's; in other words, we must try to "follow in the footsteps of the past land surveyor." When we resort to the use of distance, we assume that insufficient monuments were found. Now the surveyor must take into account the "lay of the land" or any evidence that suggests a line—perhaps a growth of pine surrounded by fields.

Direction. Did the original surveyor take a true north reading or a magnetic reading? If it was magnetic, did he ascertain that there were not metal attractions in the area that would give a false reading?

Magnetic compass headings will vary differently from true north in different parts of the country. These must be taken into account, along with normal annual changes. Our north magnetic pole lies to the west of Hudson Bay. In the eastern states, the north end of the needle will point to the west of true north, and in western states the needle will point to the east of true north. This is called the magnetic declination, and it is constantly changing. There is a point where there is zero declination. Declination values for the United States are published by the U.S. Coast and Geodetic Survey or can be found in what is known as an isogonic chart.

You can, therefore, see why it is difficult to depend upon the bearing of a line unless we are certain that it is true north or, if magnetic north, in what year it was taken. By knowing the year in which it was taken, we can update the old surveys.

Area. The least dependable factor in a deed is the area, particularly in the older deeds made prior to the use of aerial photos and tax maps. Without these, and unless the property is surveyed to some degree of accuracy, the area is usually nothing more than guesswork on the part of the owners.

Frequently, clients are advised that the area as noted in the old deeds will be less than that which is actually found when a parcel is surveyed. Perhaps this happened at a time when owners cut down on what they reported to the town in order to reduce their taxes, for today the amount of acres in old deeds will all too often be less than is noted in the new deed.

Adverse Possession

Questions invariably arise as to whether we can legally acquire land by "adverse possession" if we have had our fence on it for so long or if our driveway, which we've used for over twenty years, has been laid out over it.

Only a local attorney can advise about a particular case, but the following may give some idea of what conditions must be met to acquire title by adverse possession; habitation must be factual or true, notoriously open, hostile and belligerent, exclusive or one and only, and each of these would have to be continuous for the period of time the particular state calls for. To be more explicit:

"Habitation" or occupation requires that there be actual and visible evidence of occupation on the parcel of earth.

"Notoriously open" occupation can be some visible evidence of your use of the ground.

"Hostile and belligerent" means that there can be no agreement between the abutting owners as to ownership.

"Exclusive or one and only" means that you and only you use the parcel or section. Should there be a path or road that others might use, this would not be considered exclusive.

Occupation must be "continuous" over the period of time specified. (Here you should check with your local attorney for your particular state.)

As you can see, this adverse possession ruling is not very useful to obtain land. To prove all of these factors in a court of law would require much detailed evidence, and only a lawyer can advise you.

Attorney vs. Land Surveyor

In today's world, with all of the legal requirements needed to stand up in a court of law, the lawyer is essential to provide the client with a proper deed. But in today's world, where land is so valuable, we must also assure ourselves that a deed describes a parcel of land without any question left in anyone's mind. This can only be accomplished by the lawyer and the land surveyor working together. A lawyer cannot tell you where the land sits on the earth, and by the same token, a land surveyor cannot presume to answer any legal questions.

What Is a Title Search?

When a title search is done on a parcel of land, all records in the registry of deeds, town tax offices, and any other places where documents of land may be kept are examined to ascertain that the title to the land is marketable. A title search provides a condensed history of the ownership of a particular parcel of land.

Performed by Whom?

Title searches are normally handled by attorneys or title company employees, who have knowledge of where to look for the data that are needed. Upon completion they will either assemble a packet of documents relating to the status of the land or will submit bona fide statements to the effect.

What we want is a title which is "clear," i.e., free from any litigation, grave doubts, uncertainties, or any encumbrances in regard to ownership. If a flaw in the title is found, it must be taken care of at this point so it will not cause problems in the future. If we first purchase the land and then attempt to take out a loan to build on it, and if the bank finds a flaw that was missed, we could have a problem getting the loan.

Buyers should always inquire if a survey has already been done on the property and, if so, request a copy of it for their files.

Whatever is on file will be taken into account during these searches; however, most lawyers or title companies will provide a list of "General Exceptions"—items that they have no way of guaranteeing. A list of these would be similar to the following:

- Any conditions which would be disclosed by an inspection of the premises *and/or an accurate land survey* to provide the complete perimeter of the parcel along with its exact geographic location.
- Any persons claiming or in possession of this parcel with unrecorded leases or deeds.

- Liens that may be perfected within 120 days (varies by state law) from the last day that services are rendered, and therefore may not have been in effect at the time of this examination.
- Any encroachments, whether buildings are located entirely within the described premises, whether the roadway serving the parcel is accepted, served by town, private or otherwise, and if there is water, electricity, septic or storm drainage or any other utilities to the site.
- Any laws, codes, ordinances, or regulations affecting the use or occupancy of the premises.
- Any bankruptcy proceedings affecting the use of said property.
- Any registry errors, defects in conveyances, forgery of an instrument, rights of undisclosed heirs, or disability of grantor in his chain of title.

These list a few of the problems that can occur, even though the title to the land has been searched thoroughly by the attorney who represents you. It is well to note that he continually makes note that he cannot verify exactly where this parcel sits on the face of the earth or what its bounds are or include. These can only be answered by a proper land survey made by a registered land surveyor.

Laws relating to who can do the title search and to what extent a title search can be guaranteed should be checked with a local attorney or title-search company. They will be more informed as to the problems which can occur in any particular area.

Binder

The attorney or title-search company performing this duty will provide a so-called "binder" or document stating exactly the conditions of the deeds relating to the parcel. Depending on the area involved, there may be no need for a full title search each time a transfer takes place; here it is best to check with a local authority. Costs of title search should also be checked among the various insurance companies or attorneys.

Title Insurance and Mortgage Insurance

Title insurance protects the property owner against any flaws that may not have been found by the title searcher when the land and/or house were purchased. Before taking out title insurance, the owner should check with the attorney or title-search company to see what insurance coverage they may have. However, it is almost always necessary for the owner to purchase a title insurance policy, the cost of which is usually modest. A title insurance policy issued to a prior owner does not protect a new owner.

Nowadays, due to the numerous problems that have arisen in real estate transactions, the bank which is to hold the mortgage will take out "mortgage insurance." It should be remembered that this insurance covers only the bank, and the person buying the property must provide his or her own coverage.

Typical Title Sheet

Following is a typical request from a bank to a lawyer for a title examination and certification.

Request for Title Examination and Certification

To: Attorney ____________________ Date ____________________

From: Bank (or lending institution) ____________________

Dear Sir:

Will you please issue your title report on the property described herein. In your report, please advise of liens that will precede us and give us your opinion as to those matters which should be cleared in order to make our mortgage and the title merchantable. Describe liens by holder, amount and address of lienor. Forward the loan documents with your report, as well as your statement for services, expenses and costs rendered or to be rendered in connection with this transaction. Include costs of recording various documents, transfer tax, and issuance of title certificate.

After the mortgage has been signed and closing figures established, we will return the mortgage to you for recording. Please inform us when the mortgage deed is on file so that we may disburse.

Details of transaction:

Sales Price ______________ Refinance ______________
Name of Mortgagor(s) ______________________________
Address of Mortgagor ______________________________
Loan Amount _____ Rate _____% Monthly payment _____
Term _____ Years Type Loan Conventional _____ FHA _____
VA _____ Special Instructions ______________________
Location of Property ______________________________
Present Owner ____________________________________
Source of Title ______________ Book _______ Page _______

In the above loan, please prepare as follows:

( ) Note in duplicate
( ) Mortgage in duplicate (1 original & 1 photocopy)
( ) Warranty Deed
( ) Preliminary Examination of Title (attach map or locus when available)
( ) Extension of title
( ) Quit-Claim Deed
( ) Waiver of Lien
( ) Regulation Z Disclosure
( ) Please furnish all tax information in dollar amounts

Papers Attached:

( ) Copy of old mortgage on property
( ) Deed
( ) Abstract
( ) Note (2 each)
( ) Mortgage (2 each)
( ) Building loan agreement
( ) Disclosure statements—Regulation Z
( ) Order to pay funds
( )

Respectfully,

______________ Savings Bank

By: ______________________

Summary

In the preceding request form there is very little regarding a property survey. The major concern is to research the records of the land, and the land alone, back to a period of time that will assure there are no monetary attachments to the property. (The period of time can vary from state to state.)

Researchers will welcome a survey plan of the area if it is available, but will not require that a survey be conducted unless a definite question surfaces about boundaries. This is a chance for error, particularly since there is no research done to assure that no conflict exists in line lengths or bearings with adjacent owners' properties.

Surveyors are constantly approached by landowners or buyers requesting a survey of their property thinking that the packet of title-search material is all that is necessary as far as research is concerned. It may provide complete data on deeds for a client's parcel, but there is still the question about the deeds of all abutters. Any data that can be found in their deeds must be taken into consideration.

We must therefore consider title searches only to indicate whether a parcel of land is clear of any attachments that may fall upon the buyer's shoulders. They do nothing to guarantee the boundary lines of the land, nor do they state where such lines exist on the face of the earth unless it is expressly stated in the deeds.

Chapter 10 Assessing and Taxing Property

Truth in Phrase

We all chuckle at Ben Franklin's line, "In this world nothing can be said to be certain, except death and taxes."

Little do we realize the truth of this saying. The fact is that *taxes are the responsibility of the owner;* we must pay them whether or not we receive any bill from the town! We cannot argue in court that we did not receive notice; it is our responsibility to pay them regardless.

Furthermore, *taxes are against the property, not against the owner.* In this respect the town is covered; no matter who owns the property, taxes must be paid. Hence we find this familiar phrase in the deed, "the Grantor shall pay his proportionate share of the taxes along with the Grantee." It is usually written in relation to the period of time during the tax year that each has or will have owned the property.

State vs. Town

Taxes in most states are governed by state law. Agents of the state set the ratio of value of our property in relation to the market value. These agents are known as assessors, and they may be chosen by election or given appointments. There are also professional assessors whom towns may hire to do this work. Town selectmen may sometimes serve as assessors, but they still are governed by the state laws. Regardless of who the assessors are, it is most important that they be competent and unbiased in their work.

First Contact with Tax Office

Our first encounter with our local tax collector should be prior to the purchase of any property. Although we pay our lawyer for a title check to determine that there are no liens, mortgages, or other claims against the parcel we are interested in, there is still the possibility that unpaid taxes exist. They may have been unpaid not long enough for the town to have placed a lien against it, in which case they would not have shown up in the records the lawyer had researched. (Towns usually allow a grace period before delinquent taxes are recorded in the deeds.) The responsible person in such instances is the present owner or the owner at the instant in time when the town decides to call in unpaid taxes. Remember, taxes are against the land, and thus the present owner is responsible.

Occasionally the transaction may state that the grantor is to pay all the taxes due. But if after the sale is concluded he decides that the grantee got too good a deal, he may tear up the tax bill and forget it. Nevertheless, since the bill is against the land, the new owner owes the money to the town.

Assessing Land

Tax evaluation varies from state to state. Its major objective is to assess property at its just value, or "what a willing buyer will pay, and what a willing seller will accept." The fast-paced changes in today's real estate market have really challenged the assessors' abilities to keep abreast of them.

Population shifts from cities to the suburbs have increased the value of farmland and often contributed to the run-down condition of cities. The movement of industries has also left factory buildings to become eyesores. Then along come far-sighted landowners with plans for reviving run-down cities and others who see the factory buildings as potential apartment buildings for city workers or as potential homes for senior citizens. These changes are sudden and require constant reassessing to be certain that all property is rated in terms of "just value."

Homeowner's Right

No matter who does the assessing or re-evaluation, the owner has the right to refuse him entrance to the property. By law we can do this, and he has no comeback! Or does he?

In refusing him entry, we must consider the chance we are taking. If he cannot enter our domain to determine what we have for interior finishes, fixtures, rooms, etc., he can only guess at them, and he must be sure that he covers the overall possibilities in terms of similar houses in the general neighborhood. When his guessing is done, we can be sure that it will cover *all* possibilities, and we therefore will receive a substantial tax bill!

It is best to open the door at all times to the assessor and allow him to evaluate our homes accurately.

Check List

Most towns have a check list for evaluating houses and land for tax purposes similar to the following:

Land. Land, in most cases today, is covered by tax maps. This gives the town the area of land and location of it in respect to the town's center. Parcels of land in the areas nearer the town's center and thus nearer its utilities would be taxed at a higher rate. Farmland in the suburbs would come under another category with lower taxes, due to fewer town facilities or utilities. An owner's major concern should be that he is being taxed on his own parcel of land and for its correct number of square feet. Unless deeds are updated by a survey to determine the exact area of the land, a landowner may well be paying part of an abutter's taxes. This comes about when tax maps are made up only from aerial photos and when land boundaries are not drawn precisely on the maps.

Farmland. Even though land may be on a back road and part of a farm, assessors nowadays are likely to tax the owner for each possible house lot. Some states have laws providing for land which is in active agricultural use, or which contains a managed woodlot, to be taxed at agricultural, rather than developmental, rates. Where such protection does not exist, owners of farmland would be well-advised to have the land surveyed and soil tests for septic systems taken in order to demonstrate to the town assessors the true number of feasible house lots, as opposed to the larger number the assessors would probably assume in the absence of such information.

Buildings. Most states separate the assessments of land from those of buildings. While our bill is one figure, it reflects the total of both assessments.

Some assessments take into account the fact that the land and buildings contain a business. The value of this business must be determined as well as the fluctuations of its income from year to year.

Check List on House Tax. (See Figure 34 for typical file card.)

A normal check list for estimating your taxes on a house might be as follows:

Foundation	e.g., stone, concrete block, poured
Basement	e.g., floor, wall finishes
Framing	e.g., standard, special
Roof	e.g., wood shingles, asphalt
Interior	e.g., (by room) walls, ceiling
Exterior	e.g., brick, clapboards, shingles
Floors	e.g., oak flooring, tile, covered
Heating Plant	e.g., hot air, hot water, solar
Plumbing	e.g., copper piping, lead
Lighting	e.g., fixtures

Town Name
Property Assessment Record

Lot No.______ Card No.______ Owner______________ Date______ Book______ Page______

Date______ Book______ Page______

Date______ Book______ Page______

Date______ Book______ Page______

Street Name____________________________

Land area

Depth____________________
Width____________________
Average width________________

Square footage________________

Building value________________

Land value________________

Change values________________

Building construction

Basement walls________________
floor________________
Main floor rooms
Main bedroom________________

Kitchen________________

Interior walls________________
floors________________
Erected Date________ Add'ns________

Figure 34 *Typical File Card for Tax Purposes.*

Adjustments are then made for fireplaces, extra baths, etc. As the assesor goes through a home, each of the above items will be assessed individually. The assessor might grade them on the basis of 1 through 5, with 5 being the highest and 1 being the lowest grade possible.

With these data as well as the average factor of 1 to 5 and the square footage of the house, the assessor will refer to a table of "current-day costs of a home of this calibre." A factor of 4, for instance, might classify the home as B-C property, and if it has 800 square feet, it can easily be determined what a home of that size and classification is selling for on today's market.

Summary

What does this all mean as far as the property owner is concerned?

It means the owner should keep in close contact with the assessor and cooperate with him in determining the value of the property. With the constant changes in today's real estate market, it becomes more and more important that the property owner have precise data as to boundaries and property values. History has shown many times over that old, outdated deeds have cost many a property owner excess taxes over the years because of failure to have the property surveyed and the deeds updated with current and precise data.

Chapter 11 *Mortgage Inspections*

Due to the problems which have arisen from vague deed descriptions, banks have implemented a low-cost procedure with land surveyors to avoid such problems. This procedure has become known as a site inspection, mortgage inspection, or mortgage survey. Unfortunately, the last term, although the least accurate, has become the most popular and is misleading to the public.

The idea behind these inspections is good, and they have fulfilled a long-standing need. When these inspections were initiated, it was felt that the professional land surveyor was best suited to perform them. Knowledgeable about boundary problems, the surveyor could best point out variances between deed descriptions and actual parcel boundaries. These inspections have turned up many problems which would otherwise have gone unnoticed. They are, however, designed strictly to satisfy the banks' requirements and usually do not do much beyond confirming that the house is situated on the land in question. If their limitations are not recognized, these inspections can contain pitfalls for landowners.

Note the price of a mortgage inspection. Usually the bank arranges for it to be performed for a fee of $100 to $200. For this price, the surveyor is expected to hire an assistant, go to the site with deed in hand, take measurements of the house and any other buildings, locate the septic system and the well or town water supply, and note the electrical wires from the road to the house. Boundaries are to be noted and distances measured from them to the house and any other buildings and facilities to ensure that they are all within the parcel of land described in the deed. The surveyor then takes all this information to his office, draws up a plan incorporating it, stamps it with his registration seal, and provides copies for the bank, attorneys, and the client.

All of this has still left out two vital elements which have been emphasized throughout this book: research and contact with all abutters involved! Unless these have been done, a survey of boundaries has not been performed in a satisfactory and complete manner and cannot definitively confirm the lines.

If we lived in a world where all property was satisfactorily monumented or had been accurately surveyed previously and recorded there would be no problem. Since that is not normally the case, all property must be properly surveyed as described in previous chapters. To do a complete and definitive survey necessitates proper research, contact with abutters, field measuring and proper mathematical closures, all of which can only be done at a price.

Unfortunately, the land buyer is often misled to believe that such a mortgage inspection is a proper and definitive survey and that his property ownership is fully protected by the bare-bones procedure required to obtain a loan from the lending institution. A complete property survey appropriate to the value of the land should be obtained. It is ironic that when we purchase an item such as a $30 toaster, we tend to look more carefully at the guarantee than we do when we spend $30,000 or more for a parcel of land. The owner or purchaser must take the initiative and ensure that the land and its boundaries are adequately researched and surveyed and not leave this vital matter up to the bank or attorneys.

Chapter 12 *Surveying—Yesterday, Today, and Tomorrow*

Land Surveyor

One of our dictionaries describes the land surveyor as "a person who makes surveys to determine the area of a portion of the earth's surface by determining boundary line lengths and bearings along with the contour of the surface."

Technological advances during the past few years have increased the surveyor's capacity to provide these services with greater and greater precision. Unfortunately, the public is not always aware of the need of surveyors to update methods and techniques, leaving the surveyor misunderstood, with reactions such as, "All I need is that one line run," or "Why does it cost so much just to come out and measure one line?," or if the job is done, "Oh my gosh, is that the shape of my land?," or "Boy, you can bet the next parcel of land I buy will have been surveyed!"

To understand where surveying is today, we should first look back to its origin and trace its development to the present. Maybe then we can better understand why owners should update deeds before running into trouble.

Origin

Man's initial attempts to perform what we think of as a land survey were likely done by Egyptians. Pharaohs doled out portions of land to overseers who collected the taxes from the land, and Egyptian tombs depict surveyors in the act of measuring with a type of cord with knots in it.

The instrument used to survey this land was presumed to have been developed by the Egyptians and was known as the groma. See Figure 35. Since this is a simple instrument used to develop right angles, it is assumed by some historians that this could have been developed by other peoples at about the same time.

The Egyptian's primary use of the groma was for the replacement of lines after the Nile River had overflowed its banks and obliterated boundary lines.

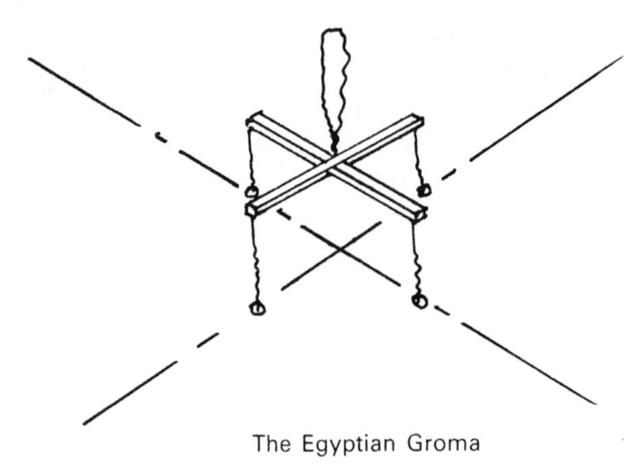

Figure 35 *Instrument used by Egyptians for the layout of long lines at right angles.*

Geometry

The understanding of mathematics brought forth further developments in the use of geometry to allow the surveyor to compute distances by use of protractors to determine the angles and of tapes for measuring the distance. With two angles and the included side, two sides and the included angle, or with all sides, it was possible to compute the unknown portions of the triangle. See Figure 36 for the use of the protractor by the land surveyor. This was called triangulation, and it plays a major role in property surveys today.

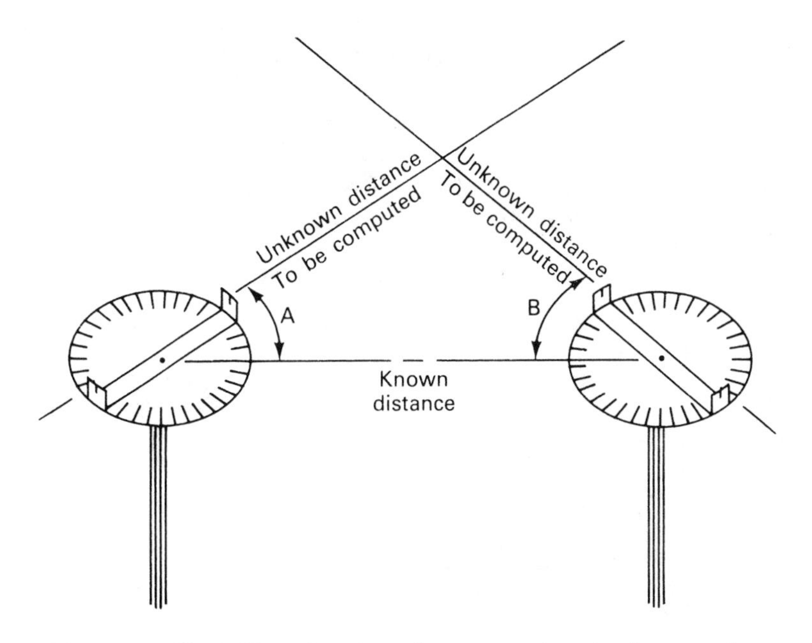

Figure 36 *Simple illustration of use of a protractor for estimating an angle so the distance can be computed.*

Chinese Contribution

Chinese development of the principle of magnetized needles floating in a fluid to point to magnetic north was adapted to use in conjunction with the protractors, and simply put, the telescope was then added to produce what is known as a transit. See Figure 38A. Improvements, such as the vernier to read the angles down to seconds and the levels for setting the instrument up, were then added as time went on.

A simpler unit was used when man incorporated the compass with a couple of upright slit bars through which he could sight short distances and project a straight line by bearings as in Figure 37. The compass, along with the measuring chain, was used in the early period of our country to divide it into settlements. This chain (Gunter's Chain) was usually 66 feet or four rods long and consisted of 100 links of 7.92 inches each. Many of today's deeds still refer to this method of measurement when they describe metes and bounds, and show the antiquity of the deeds being used. Many old-time rail fences consisted of 11-foot sections, which permitted them to be used to measure out 66 feet—or one Gunter's Chain of four rods.

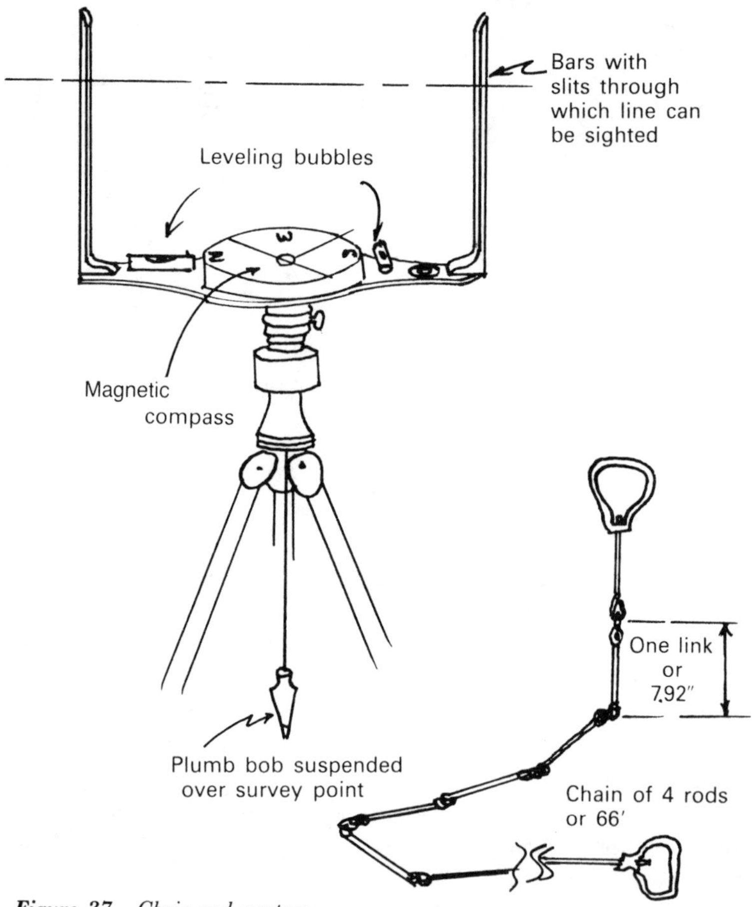

Figure 37 *Chain and compass.*

Transit and Tape

The next step in the revolution of instruments was use of the transit and tape for greater accuracy. With the transit it was possible to read the angles of the turns in the lines being surveyed instead of reading the compass headings. See Figure 39.

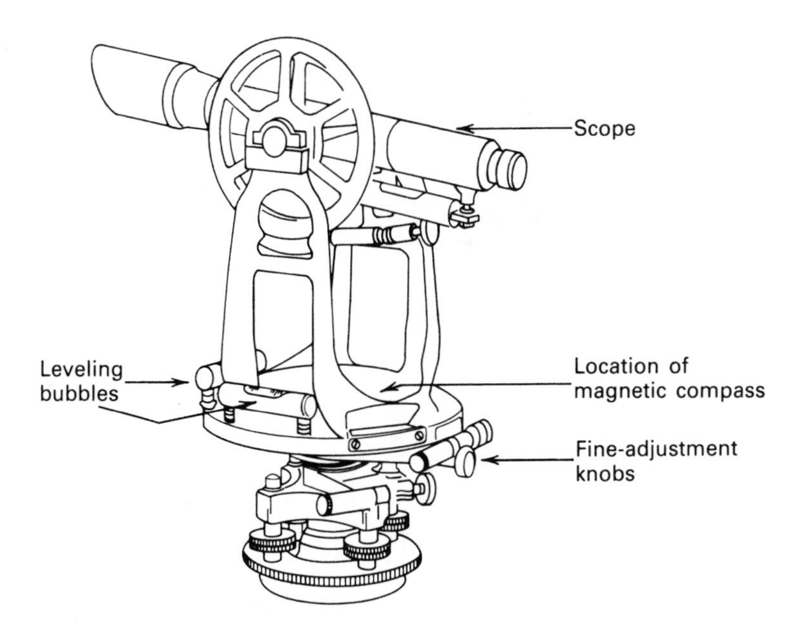

Figure 38A *Example of engineer's transit.*

Now the surveyor could turn the angles in degrees, minutes and seconds in order to assure that the parcel he was working did "close," i.e., angles totaled 360 degrees. This provided the surveyor with better figures of the angles than could be obtained from compass headings, which were prone to errors caused by magnetic interference and were unreliable.

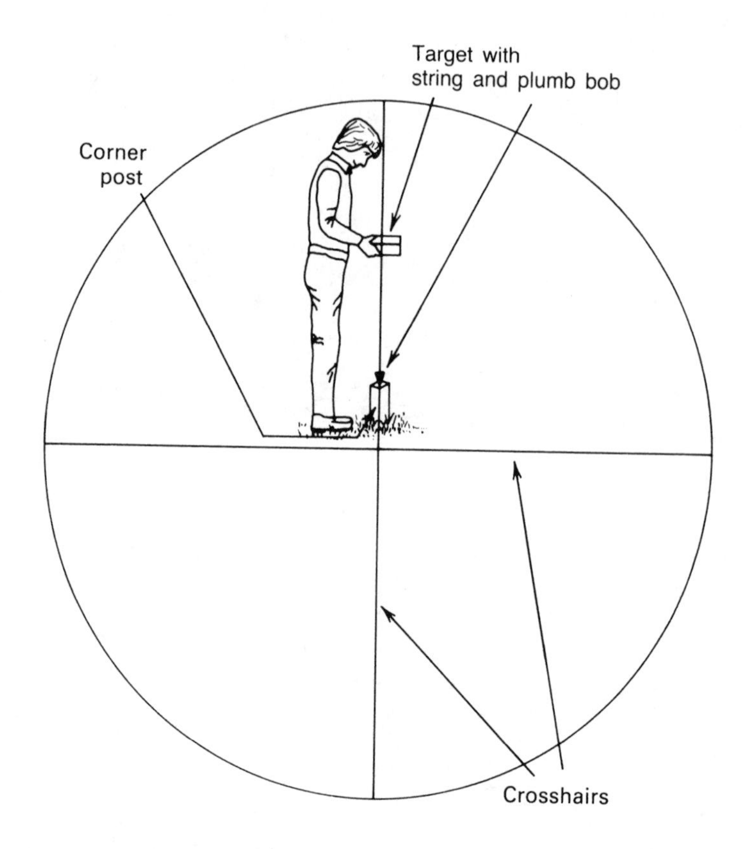

Figure 38B *Typical view through the scope of the transit which is being aligned on the target and plum bob suspended over the corner marker. The vertical line is aligned for accuracy, and readings are taken on the transit's table for magnetic headings and angles between corners.*

Other improvements were later made on the instruments, such as the self-leveling transit and optical plummet to assure that the transit was centered over the point even in a windstorm.

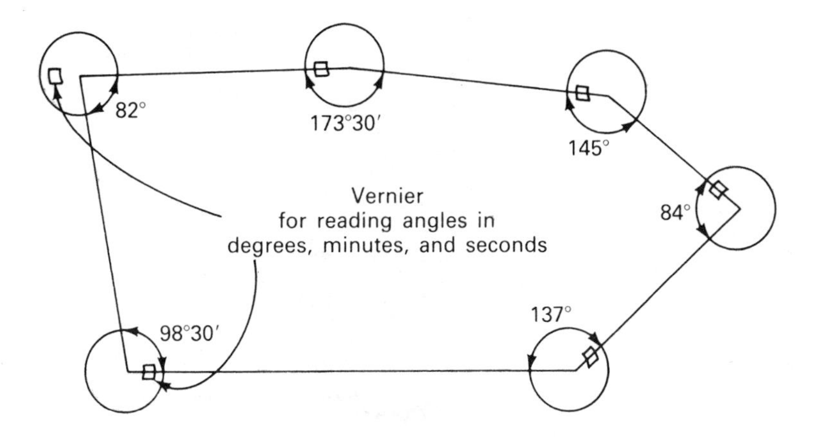

Figure 39 *With the addition of a Vernier scale, it became possible to subdivide angles into still smaller settings of minutes and seconds and to achieve an accuracy that would have been impossible without it.*

Recent Survey Developments

During the Second World War, the use of aerial photos to discover enemy positions brought about advancements in techniques that became useful in mapping and in doing land surveys. Later, towns discovered their use for tax maps.

A major development came about when the electronic measuring device was invented. It is known as the EDM—the electronic distance measurer. See Figures 40A and 40B. By the flick of a switch, the distance between two points can now be determined with accuracy to a thousandth of a foot. It is now no longer necessary to measure a line with a 100-foot tape, taking errors into account for sag, temperature, expansion, etc.

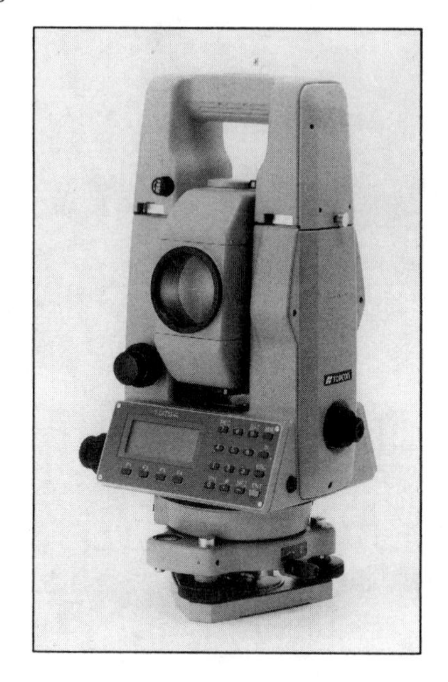

Figure 40A *A Total Station, which combines a transit and electronic measuring device.*

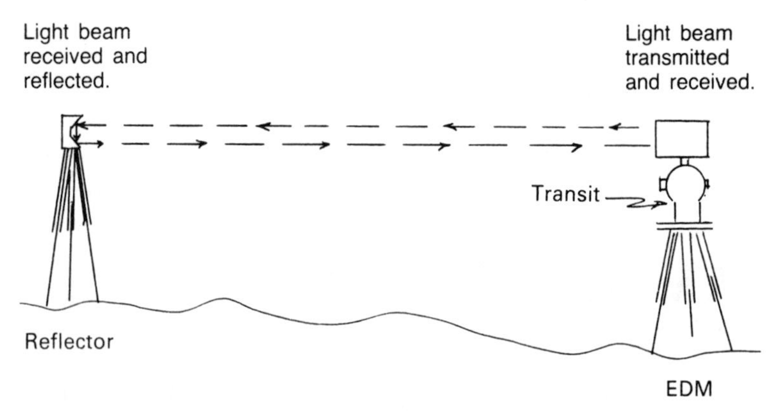

Figure 40B *Simple characteristics of Electronic Distance Measuring (EDM).*

Calculators and Computers

A still more sophisticated development available to the surveyor is a complete computer system. One of its components is a field storage unit for data obtained from the electronic distance-measuring device when working in the field. When taken back into the office, it is connected to a survey computer which formats the information received, computes the closed survey for the parcel of land, and forwards this information to another machine which draws the plan of the area. See Figure 41.

Geographic Information System (GIS)

Most of us have seen pictures of the medical equipment that doctors use to scan the brain. A person is shown lying on a platform, head projecting inside a ring. As the person is moved farther and farther into the ring, different layers of the brain are exposed to assist the doctor in diagnosing the patient's condition correctly. Using the capabilities of modern computers and available data, layer upon layer is exposed and depicted. This is basically what the computer mapping system known as GIS can do for us.

Do you need a map showing roads, sewer lines, and soil types? Or a map showing all landowners abutting a proposed highway or other project? A map showing all parcels of land greater than 100 acres and their owners? Need to see where elderly people living on back roads are located, so that they can be reached quickly in the event of a medical emergency, storm, or disaster? Or a map showing the location and types of agricultural land?

These are just a few examples of what GIS can do, provided the right information has been entered in the computer.

How GIS Works. The GIS process usually starts with the computer tracing features from aerial photos. It usually takes about

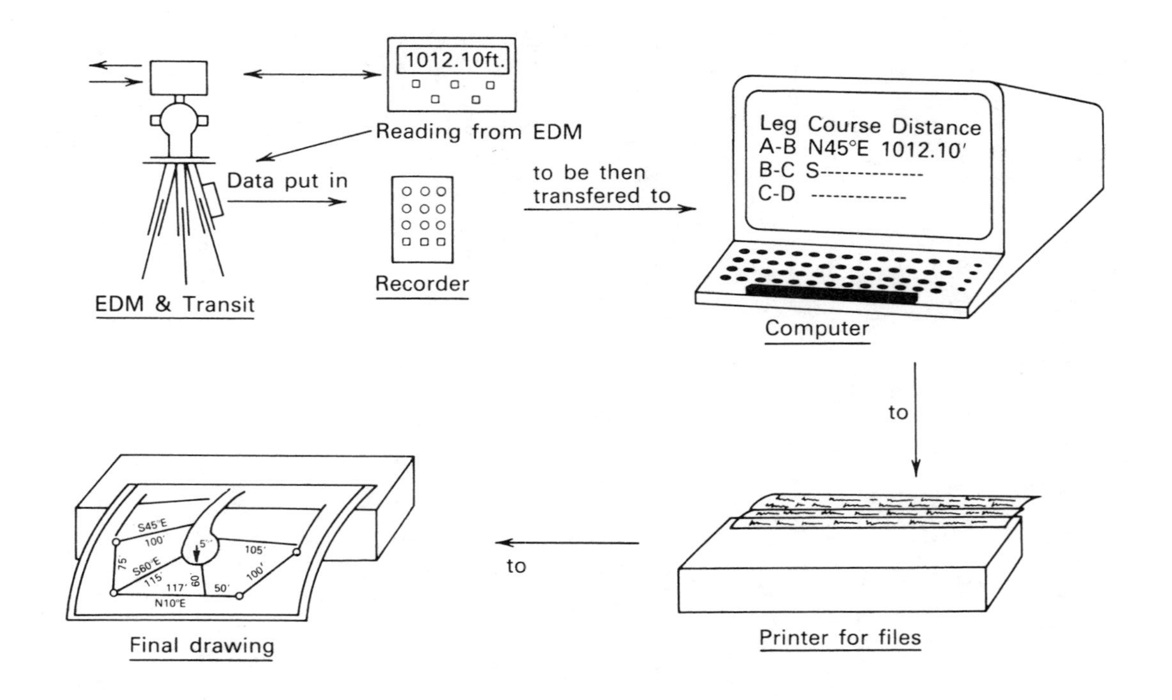

Figure 41 *Sample configuration of the total survey from field to drawing.*

ten aerial photos to cover the area of a typical New England township (although the number can vary depending on the density of houselots). These traced maps can then be combined with any number of layers and sets of data stored in the computer. There could, for instance, be a layer showing roads and data on road numbers, surface type, width, or speed limits. Or one showing water, sewer, electrical, and telephone lines. Another could show soil types by parcel, together with information on acreage, slope steepness, suitability for septic systems, agricultural potential, or depth to bedrock.

The computer can resize its collection of maps to whatever area is needed, print out that area on any size sheet, and display whatever combination of layers and data is needed. For clarity, each feature can be displayed in a different pattern or color.

The Uses of GIS. The uses of GIS are almost limitless. Listers can use GIS to call up information on names and addresses of property owners, when properties were assessed, taxes levied, and whether taxes have been paid. Members of planning and zoning boards will find GIS's instant information on utility lines, soil types, flood plain locations, etc., of great assistance in considering applications for building permits, subdivisions, and other projects. The uses for highway departments, fire and ambulance services, and construction and utility companies are obvious. GIS is also increasingly important to towns and regional bodies attempting to coordinate such projects as inter-town roads and developments which straddle town lines or impact on neighboring towns. Finally, you as an individual landowner can learn a lot about your own property from GIS.

There is nothing mysterious about GIS. It simply performs in minutes mapping and data gathering tasks that in the past would have taken us days and weeks to do. It has come at the right time. With today's increasingly complex planning and zoning requirements, towns and citizens need more types of map information more quickly than every before. GIS can help towns to grow and build in ways beneficial to all their citizens.

GIS and You. As in all other computer-related fields, equipment and software for GIS is becoming more and more affordable. Increasingly, towns, counties, or other regional bodies have GIS equipment and personnel trained in its use. If you are involved in your town's planning, zoning, or tax planning activities, you may be able to get together a group of townspeople who work with maps and have your state or local government give you a demonstration of how GIS works and how you can work with it and benefit from it.

Surveying by Satellite (Global Positioning System)

The increase in land values has created a demand for more precise ways of locating parcels "on the face of the earth," which, as noted in Chapter 2, can be difficult if there are no permanent markers to which the description of the land can be tied. Once again, the civilian industry has benefited from the results of military research. The Global Positioning System, or GPS as it is known, was designed primarily as a navigational and targeting system. Precision was important to our military forces to assure that they could hit a military installation and not the nearby civilian population.

Accuracy. Combining the techniques of conventional surveying with "satellite surveying" gives us a higher degree of accuracy. With GPS, it is conceivable that our houses could in the future be referenced by coordinates (similar to the latitude and longitude lines on a map), eliminating the need for a boundary survey.

A surveyor using GPS sets up instruments capable of receiving signals from orbiting satellites. Two, three, or more of the satellites currently in operation pass over his region of the earth. His instruments receive their signals, and the location of each instrument is then plotted to a high degree of accuracy through a process similar to the triangulation described in Chapter 2.

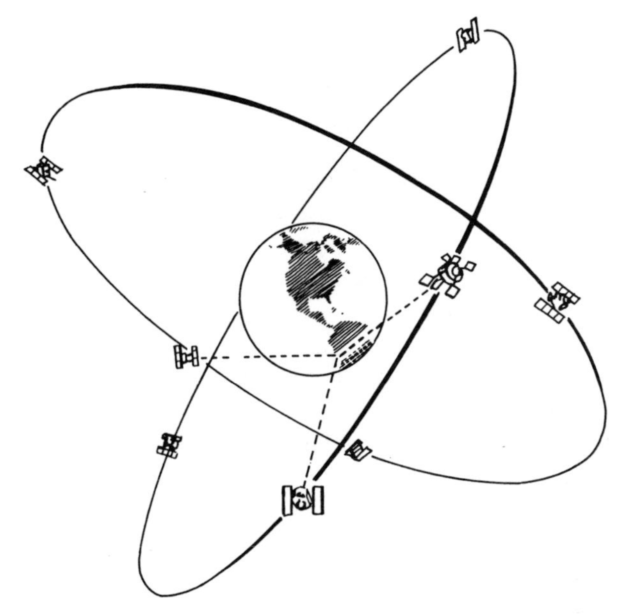

Figure 42 *Surveying by satellite.*

Instruments for receiving GPS satellite signals and performing the necessary calculations are available, and, as is usually the case where new technology is concerned, are gradually becoming cheaper to manufacture. There are still some minor details to be worked out, but as GPS equipment becomes more affordable, the surveyor should be able to make greater use of this method in the future.

Benefits of GPS. One of the benefits of GPS is that it is all-weather. A crew does not have to be exposed to the elements for long periods of time. GPS can be used independently of a line of sight between two points and is available throughout the year, day or night. GPS can also be used by aircraft flying over towns making aerial photos.

The accuracy of tax maps can also be improved by using GPS. With the accuracy and flexibility which GPS offers, it becomes easier to show the inner sections of town, where greater detail is needed, in larger scale, while a smaller scale is used for the more sparsely settled outer areas.

More precise depictions of road layouts on highway maps are also possible with GPS. A surveyor with GPS instruments in his car can drive along roads and stop to take readings at as many points or intersections as necessary.

Cost. Currently GPS is too costly to be of practical use to the individual landowner and is cost-effective only for large surveys. However, as technology advances, the price of equipment falls, and more professional land surveyors receive training in the use of GPS, the benefits of satellite surveying will become more broadly available and more affordable.

The Professional Surveyor

All U.S. states now require that large property surveys be performed by registered land surveyors, but some states exempt smaller jobs from this requirement. Hence it is advisable to check that a surveyor is registered in the state where the job is. "Weekend" surveyors abound, and since their clients usually have no knowledge of what is necessary for a complete survey, many surveys are incomplete. Although lines are furnished in the field and drawings provided, such surveys run the risk of future challenge.

The professional surveyor to be recommended is the one who works at his trade constantly, maintains his equipment, follows set standards, and keeps up with all regulations set by his local society of land surveyors. His credentials can be checked by asking for his state registration certificate or card, and the registration card of the state surveyors' society. A surveyor should also belong to the American Congress on Surveying and Mapping, which keeps him informed of relevant national news.

Land Survey Costs

As in anything else, it is important to have some idea what the survey is going to cost to have done and have done right. There are three factors to consider: research, field survey, and final drawing. To ask for an estimate of what a survey job will cost is something akin to asking a lawyer what it is going to cost for him to defend you in a lawsuit. Granted, he can give you a general idea, but to quote an exact figure at the outset is seldom possible. So too with a survey job. Today's high land values require that surveyors not only measure what the client feels he owns, but also ascertain that all abutters are contacted. And when the final drawing is done, the surveyor must certify that all possible sources of information has been covered.

By talking with a surveyor, it is possible to get a general idea what his average costs run to do a lot, or perhaps he can quote you a per lineal-foot price if your land is more than a house lot in size.

It is most important to know what is provided. Today's land surveyor should be providing a drawing that can be filed in the registry of deeds, and on that plan should be all information that will enable you to have this parcel reproduced without challenge on the face of the earth.

If a surveyor has done the job properly, a complete data file will be available should future questions arise. And finally, all questions that could possibly arise with any abutters will have been cleared up and they will have been provided with plans of the survey to keep with their deeds for reference if their land is subsequently surveyed.

Appendix A *Glossary*

Abutter One whose property abuts, is contiguous, or joins at a border or boundary.

Acre A quantity of land containing 160 square rods, 4,480 square yards, or 43,560 square feet, in whatever shape.

Adverse Possession A method of acquisition of land by possession for a statutory period of time under certain conditions.

Area Coordinates A system of laying out a survey on the x axis or east-west and y axis or north-south.

Bearing The compass direction or heading of one object or point from another point.

Bench Mark A permanent point the elevation of which above some definite or assumed datum is known.

Bounds The external lines, or limiting lines, of property.

Boundary Every separation, natural or artificial, which marks the lines of division of two adjoining properties.

Chain Refers to either the engineer's or Gunter's (surveyor's) chain. The Gunter's chain is 66 feet (4 rods) long.

Chain of Title Successive conveyances, or deeds, affecting a parcel of land.

Clear Title Good title of deed, clear from any defects or limitations.

Cloud on Title A claim or encumbrance which could affect or impair the ownership of a particular parcel.

Control Points Points that have been established in a series with a high degree of accuracy or precision, to which parcels may be tied.

Course The direction of a line with reference to the meridian.

Declination The angle between magnetic north and true north.

Deed A document or writing signed by a grantor which conveys land from one party to another.

Departure With the north-south line being the latitude of a survey, the east-west line is called the departure.

Easement Right of use over the property of another.

EDM (Electronic Distance Measurer) A device which, by registering the time it takes for a beam of light to travel between two points, makes it possible to measure distance with great accuracy.

Eminent Domain The power to take private property for use where the public interest can best be served.

Fee Simple Fee Simple is absolute ownership (of land) with unrestricted rights of disposition.

Filed Plan Drawing of plan, showing all vital data gained during a survey, which has been filed in the registry of deeds.

Geodetic Marker A survey marker placed by the government in either a horizontal or vertical plan.

Grantee One to whom a deed or grant is made.

Grantor One from whom a deed or grant is made.

Hectare Metric unit of measure denoting 10,000 square meters (2.471 acres).

Highway Marker A stone bound placed by the State Highway Department denoting the sidelines, curve points, and offsets of highway points.

Latitude The distance of a point on the earth's surface from the equator.

Linen Type of material upon which ink drawings of land surveys have traditionally been drawn for filing in registries of deeds. Nowadays, Mylar is more commonly used, but the term "linen" is often still used to describe the durable copy filed with the registry.

Longitude Distance on the earth's surface east or west from a meridian point.

Metes and Bounds The boundary lines of land, with their terminal points and angles. Described by compass headings and distances.

Minute The sixtieth part of a degree, as of longitude.

Monument Point or mark indicating lines on boundaries of a land survey.

Mylar Clear plastic material on which plans are drawn for filing in registries of deeds.

Natural Monument Streams, lakes, shores, ledge outcrops, and sometimes streets and highways, used as boundary markers.

Parcel A part or portion of land.

Quit-claim The release or relinquishment of a claim to land.

Quiet Title Action To bring into court a proceeding to establish title to land by forcing the claimant to establish his claim.

Pole Unit of measure—25 links or 16 1/2 feet.

Registry of Deeds Place where deeds and plans of land, mortgages, and realty instruments are filed.

Right of Way The right of a person to pass over the land of another.

Riparian Belonging to or relating to the bank of a river.

Rood One quarter acre or 10,890 square feet.

State Acre 40,000 square feet. A measurement increasingly used by state and local governments, due to the ease of performing calculations with round numbers.

Survey Process by which land is measured for ascertaining its boundaries.

Theodolite A term often used synonymously with "transit" (see below). The modern theodolite usually has a more powerful telescope than a conventional transit and includes a built-in level. A theodolite can also give digital readings in degrees, minutes and seconds, to provide interior angles for all corners of a property. A bearing (either magnetic or true) is taken on one line of the property, and then all other line bearings are computed from these corner angles.

Transit A sighting instrument used for measuring horizontal, and sometimes vertical, angles. In today's transits, sighting is usually done with a telescope, which revolves on a flat metal disc divided into degrees and minutes. For greater accuracy, a vernier scale normally supplements the disc to provide readings in seconds to the nearest 30, 20, 10 or lower, depending on the precision of the equipment. A transit also has a compass which gives a reading in magnetic degrees. See Figure 38A.

Traverse Series of bearings and distances from one point to another.

Title Search Process by which records in the registry of deeds are examined to determine proper ownership of a parcel of land.

Vernier Scale A short graduated scale that slides along a longer graduated scale and allows the longer one to be subdivided into smaller parts. Used in surveying to divide angles into minutes and seconds for great accuracy.

Warranty Deed A deed by which one binds oneself by covenant to defend the grantee in his title and possession.

Appendix **B** *Conversions*

1 link	– 7.92 inches
25 links	– 1 rod, perch, or pole
1 rod	– $16^1/_2$ feet
4 rods	– 1 chain (Gunter's), 66 feet or 100 links
1 rood	– $^1/_4$ acre (or 104.355 feet × 104.355 feet if square) or 10,890 square feet
1 acre	– 160 square rods or 10 square chains
1 acre	– 43,560 square feet (or 208.71′ × 208.71′ if square)
1 state acre	– 40,000 square feet
640 acres	– 1 square mile or section
36 square miles	– 1 township, 480 chains square, or about "23,000 acres"

Example conversion: to convert 14 rods 12 links into feet.

14 × $16^1/_2$ feet	....	231.00 feet
12 × 7.92 inches / 12 (inches per foot)	= 7 feet 11.04 inches or	7.92 feet
	Total	238.92 feet

1 meter	– 39.37 inches or 3.2808 feet
1 kilometer	– 1000 meters
1 hectare	– 2.471 acres or 1 square kilometer

1 vara	– 33 inches in Spanish, Mexican, Brazilian, Californian, or South Western states.
1 Texas vara	– $33^1/_3$ inches, adopted in 1919.
1 arpent	– approx. 0.85 acres (used in grants of French-crown – varies with different states.)

Appendix C *Landowner's Checklists*

Items to check prior to purchase of land:

1. Has the property been surveyed? If not, are there defined boundaries not questioned by the abutters? Look at all corners when they are clear and not covered with snow, which can hide boundary markers.

2. If the property has been surveyed, is that previous work recent enough and adequate for your purposes? As noted in Chapter 2, changes in the location of magnetic north can invalidate old instrument readings. Or, you may be considering a parcel comprising several previously surveyed lots. The previous surveys may have employed different methods of varying precision. Some may have been merely "acre surveys," intended to do no more than measure the land and calculate the number of acres. In such a situation, it is important to do a survey on the perimeter of the combined parcels.

3. Is a public sewage line available to the land? If not, has a soils test been made, to determine whether the town will approve a septic system on the land? It is advisable to check this before spending money to survey the land.

4. What sources of water are there? Is town water available, or will a well have to be drilled?

5. If there is already a well on the property, how far is it from the septic system on the adjacent lot? Some states have laws describing how far a septic system must be from a well but do not regulate how far from an existing septic system a new well must be drilled. If the well is near a road, could there be problems with road salt?

6. If you are buying land with a building on it, do you have plans for altering the building's size or appearance? If so, make sure that town zoning and design ordinances will permit the changes you hope to make. Be cautious about using a real estate company drawing as a basis for these judgments, as such drawings may not show exactly how the building is located on the land or depict the shape of the land accurately. (See Figure 24).

Assignments For Your Surveyor

*1. **Research.*** View the general area to be surveyed and note all corners which are marked or need to be marked. Obtain all available data on the property to be surveyed and on abutting parcels from town, county, or state records and from any other sources of information. Discuss the survey with all abutting landowners and obtain whatever pertinent data they can offer. Contact all other persons who may have useful knowledge of the area.
*2. **Preliminary Cost Information.*** Upon completion of this research prepare an estimated cost for performing the survey, with a detailed breakdown of the costs of all assignments that will be covered in the survey.
*3. **Field Survey.*** Perform a survey of the property consistent with local codes and appropriate to the uses which will be made of the final plan and consistent with local codes. After evaluation and interpretation of all data, set markers of a type specified by the owner or by local codes at all required locations.

*4. **Drawing.*** Provide a scaled drawing of the property survey showing all data necessary for locating this parcel of land on the face of the earth. Note all essential and identifying data, such as rights-of-way, easements, stone walls, barbed wire fences, etc. The drawing should be on Mylar or other durable material acceptable to the local registry of deeds. Enough prints should be made to meet the owner's requirements, for the surveyor's file, and for the abutting landowners (assuming they have been consulted and have agreed to the corners set).

*5. **Recording.*** When recording the drawing at the registry of deeds, the surveyor should ask the registry to stamp your copy and the surveyor's with the notation that the survey has been recorded and is on file in Plan Book ____________ Page ____________. This notation should be copied onto all prints made for the bank, lawyers, abutters, etc.

*6. **Billing.*** Final billing should show the complete breakdown of all costs involved in the survey.

Index

Abutters, 30, 108
  rights of, 70
Acre, 108
  state, 110
Acreage, inflated, 7
Adverse possession, 78-79, 108
Aerial mapping, 21-28, 99, 101-103
Angled lines, problems of, 52-56
Angle reading, 12-14
Angle, interior, test, 41-42
Area, 78
  coordinates, 60-63, 108

Base lines, 52-53
Bearing, 108
Bench mark, 108
Binder, 81
Boundaries, 108
  disputes, 76
  and fences, 74
  kept in memory, 4, 30-31
  markers, 65-69, 77
    replacing, 70-74
  problems with, 1-8
  walking the lines, 29, 66-67
Bounds, 108

Chain. *See* Measuring chain
Chinese influence on surveying, 95
Closure. *See* Mathematical closure
Compass, 76, 95-96
  headings, 13, 39, 77-78
Computers, 101-103
Control points, 108
Coordinate method of laying out parcel of land, 60-63, 108
Corner markers, 65, 77
  replacing, 70-74
Course, 108

Death of landowner, 4, 7
Declination, 108
Deeds, 108
  descriptions, 3, 9-10, 60-63
    problems with, 50-56, 76
    writing, 56-59, 75-76
  elements of, 49
  history of, 49-50
  and inherited land, 4-5
  intent, 64
  vs. land, 6
  problems with, 6-7
  quit-claim, 55, 59-60, 110
  recording vs. registering, 33
  referring to a recorded plan, 37
  registry of, 32-34, 110
  unregistered, 30
  warranty, 60-63, 111
  wording of, 63-64
  writing, 75-76
Departure, 109
Direction, 77-78
Distance of boundary lines, 77

Easement, 109
Egyptian influence on surveying, 93-94
Electric distance measurer (EDM), 42-45, 99-101, 109
Eminent domain, 109
Error of closure, 47-48

Fee simple, 109
Fences, and property lines, 74
Field notes, 14-15
Field survey. *See* Survey
Filed plan, 109
Fire departments, use of aerial maps, 25

Geodetic marker, 109
Geographical Information System (GIS), 101–104
  use of aerial maps, 25
Geometry, 94–95
Global Positioning System (GPS), 104–106
Grantee, 109
  index books, 34, 36
Grantor, 109
  index books, 34
Groma, 94
Gunter's Chain. *See* Measuring chain

Hectare, 109
Highway departments, use of aerial maps, 24–25, 28
Highway marker, 109
Hills, 19
Homeowner's rights and tax assessors, 87

Inherited land, 4–5, 30–31
Insurance, 82
Interior angle test, 41–42

Land
  adverse possession, 78–79, 108
  assessing, 86–88. *See also* Tax
  coordinate method of laying out parcel, 60–63, 108
  vs. deed, 6
  inflated acreage, 7
  inherited, 4–5, 30–31
  items to check prior to purchase, 113–115
  markers, 65–69, 77
    replacing, 70–74
  plan books, 34–38
  researching, 29–38
  subdivisions, 51
  unclaimed parcels, 23–24
Land surveyors. *See* Surveyors
Latitude, 109

Lawyers
  role of, 75–76, 79
  title search, 31, 80–81
Linen, 109
Lines, walking, 29, 66–67
Loggers, 31
Longitude, 109

Magnetic compass. *See* Compass
Magnetic declination, 78
Maps
  aerial, 21–28, 99, 101–103
  tax, *vii,* 21–24, 26–27
Markers, 65–69, 77
  replacing, 70–74
Mathematical closure, 39–48
  computations, 41–44
  degree of, 47–48
  inaccuracies, reasons for, 42–46
  interior angle test, 41–42
Measurement conversions, 112
Measuring chain, 76, 95–96, 108
Measuring tape, 42–45, 97–98
Metes and bounds, 109
Minimal error of closure, 47–48
Minute, 109
Monument, 110. *See also* Markers
Mortgage inspections, 91–92
Mortgage insurance, 82
Mortgage surveys. *See* Mortgage inspections
Mylar, 110

National Geodetic Survey, 56

Offset lines, 15–18

Parcel, 110. *See also* Land
Plan, recorded, 34–38, 109
Planning boards, 4
  use of aerial maps, 28
Point of beginning, 11
Pole, 110
Population explosion, 3–4
Property tax. *See* Tax

main

Quiet title action, 110
Quit-claim deed, 55, 59-60, 110. *See also* Deeds

Range lines, 52-53
Real estate brokers, use of aerial maps, 25
Recorded deeds, 33
Recorded plans, 34-38
Registry of deeds, 32-35, 110. *See also* Deeds
Right of way, 110
Riparian, 110
Roads, angled, 52-56
Rood, 110
Running lines, 15-19

Site inspections. *See* Mortgage inspections
State acre, 110
Surveying
- accuracy vs. precision, 46-48
- angled lines, 52-55
- costs, 107
- history of, 1-3, 93-99
- horizontal measurements, 19
- inaccuracies, reasons for, 42-46
- legal sequence in, 77-78
- locating on the face of the earth, 20
- method of, 9-20
- and obstructions, 16-19
- by satellite, 104-106

Surveyors
- checking credentials, 106
- use of aerial maps, 25

Surveys, 110
- old and imprecise, 76
- recorded plans, 34-38
- unregistered, 27, 30

Tape. *See* Measuring tape
Tax, 85-90
- assessors, 85, 87
- evaluation, 86-87
  - check list, 87-90
- maps, *vii,* 21-24, 26-27

Theodolite, 42-45, 110
Title
- chain of, 108
- clear, 108
- insurance, 82
- search, 31, 80-81, 111
- sheet, 82-83

Town boards, use of aerial maps, 25
Transit, 11-12, 42-45, 97-98, 100, 110-111
Traverse, 111
Triangulation, 16-19, 94-95

U.S. Department of Agriculture, 23
U.S. Department of Commerce, 56
Utility poles, 69

Valleys, 19
Vernier scale, 99, 111

Warranty deed, 60-63, 111. *See also* Deeds